TARGETING AMERICANS

TARGETING AMERICANS

The Constitutionality of the U.S. Drone War

H. Jefferson Powell

Foreword by

Philip Bobbitt

OXFORD
UNIVERSITY PRESS

OXFORD
UNIVERSITY PRESS

Oxford University Press is a department of the University of Oxford. It furthers the University's objective of excellence in research, scholarship, and education by publishing worldwide.

Oxford New York
Auckland Cape Town Dar es Salaam Hong Kong Karachi Kuala Lumpur Madrid Melbourne Mexico City Nairobi New Delhi Shanghai Taipei Toronto

With offices in
Argentina Austria Brazil Chile Czech Republic France Greece Guatemala Hungary Italy Japan Poland Portugal Singapore South Korea Switzerland Thailand Turkey Ukraine Vietnam

Oxford is a registered trademark of Oxford University Press in the UK and certain other countries.

Published in the United States of America by
Oxford University Press
198 Madison Avenue, New York, NY 10016

Library of Congress Cataloging-in-Publication Data

Names: Powell, Jefferson, 1954- author.
Title: Targeting Americans : the constitutionality of the U.S. Drone War / H. Jefferson Powell; foreword by Philip Bobbitt.
Description: New York : Oxford University Press, 2015. | Includes bibliographical references and index.
Identifiers: LCCN 2015037930 | ISBN 9780190492847 (hardback)
Subjects: LCSH: War and emergency powers—United States. | Targeted killing—United States. | Constitutional law—United States. | National security—Law and legislation—United States. | State crimes—Law and legislation—United States. | Drone aircraft—Law and legislation—United States. | Targeted killing—Moral and ethical aspects. | Counterinsurgency—United States.
Classification: LCC KF5060 .P695 2016 | DDC 343.73/015422—dc23 LC record available at http://lccn.loc.gov/2015037930

1 3 5 7 9 8 6 4 2

Note to Readers

This publication is designed to provide accurate and authoritative information in regard to the subject matter covered. It is based upon sources believed to be accurate and reliable and is intended to be current as of the time it was written. It is sold with the understanding that the publisher is not engaged in rendering legal, accounting, or other professional services. If legal advice or other expert assistance is required, the services of a competent professional person should be sought. Also, to confirm that the information has not been affected or changed by recent developments, traditional legal research techniques should be used, including checking primary sources where appropriate.

(Based on the Declaration of Principles jointly adopted by a Committee of the American Bar Association and a Committee of Publishers and Associations.)

To David,

amico carissimo

CONTENTS

FOREWORD

In every state of which the international system is composed, the constitution is necessarily involved in the making and execution of the state's strategy. The nature of that involvement is one dimension by which we determine the character of a particular state. The subordination of the professional military to elected representatives of the state; the making of legal regulations governing land and naval forces by the lawmaking body; the fashioning of rules of engagement by an elected executive; and above all, the parliamentary control of the decision to go to war that characterize states of consent—which in the twentieth and twenty-first centuries for the most part meant the constitutional democracies. At present, however, a combination of many historic factors is leading us to a moment of crisis in the relationship between law and strategy, and that crisis cannot be avoided by those democracies.

It is a source of some disquiet that we have wasted so much time avoiding addressing this crisis, for appearances to the contrary notwithstanding, we are living in a relatively tranquil period. This is the time we should begin the painstaking legal and strategic analysis that will preserve our constitutional institutions. Once we are compelled to innovate in the face of mass destruction and before the threat of historic annihilations, we run the risk that we will undermine our own institutions in our panic and our haste. It is an unpleasant irony that one of the principal reasons we have not undertaken more searching assessments of the reform of law and the refashioning of strategy is the chorus of voices shouting that our actions since September 11, 2001, have already utterly defaced the U.S. Constitution. In such a hyped-up drama, who would dare to undertake a patient and careful legal analysis?

H. Jefferson Powell, in this rewarding and thoughtful book, has done just that. Here, Powell addresses the vexing question of targeted killing from a constitutional point of view. The fundamental issue is whether the Authorization for the Use of Military Force (AUMF) adopted by the Congress in 2001 provides lawful authority for the president to single out a person or group for attack by U.S. forces when the target consists of individuals, some of whom may be American citizens, who have not been tried and convicted for a capital crime.

This question is sometimes confounded by other questions like: whether the United States needs a declaration of war rather than a joint resolution like the AUMF; whether the AUMF of 2001 can still be in effect more than a dozen years later when the object of congressional retaliation has been decimated and yet the end of the conflict known as the "wars on terror" is not in sight; and whether the ban on assassinations memorialized in Executive Order 12333 or perhaps the customary international law of some centuries barring assassinations doesn't bar targeted killing, too. Powell handles these ancillary issues with a masterful touch.

Along the way, there are many points as to which in my mind there remain dispute. Does the Fifth Amendment's Due Process Clause apply to targeted killing? (I think it does; Powell disagrees.)

Does congressional action, whether by declaration of war or by legislative action like the AUMF reduce the president's Article II authorities, as might be suggested by Justice Jackson's concurrence in the *Youngstown* case, which arose in the context of the Korean War? (I doubt that it does in the present circumstances.) Do ethical arguments—one of the modalities of constitutional argument that students are taught in the Standard Model—have any role in this debate? (On this point, Powell and I emphatically agree on the important salience of the role of ethos.)

A further question is raised by a memorandum authored by William Rehnquist, when he was the head of the Office of Legal Council, addressing the question whether hostilities authorized by Congress must be directed at that "nation against whom war has been declared." This is highly relevant today when al-Qaeda, against whom the AUMF was directed, has spawned al-Qaeda in the Arabian Peninsula (AQAP), which in turned has morphed into the Islamic State (ISIS). In fact, there are ample precedents for using congressional authorizations for war to apply to a subsequent group when hostilities emerged from the conflict initially authorized. The most pertinent is the Philippines War of 1898 in which U.S. forces, having defeated the Spanish against whom Congress had authorized war, were confronted by Philippine insurgents who had actually fought the Spanish. United States forces proceeded to wage war against the Philippine insurgents without any further congressional authorization, choosing instead to rely on the initial declaration of war against Spain out of which the insurgency had emerged. There are also the examples of Albania, Croatia, and Slovakia from World War II that established the principle that fresh authority is not needed if the new belligerent emerges from the originally authorized conflict.

But this book isn't intended to *end* all debate. Indeed, that's the point of a superb essay like Powell's: that it structures our analysis, provides a playing field with rules on which we can agree, narrows the points of contention, and gives us an overall assessment that is incontrovertible. Once delimited, we can argue more fruitfully. Without such an overview and without the legal methods by which it is created and maintained, we are likely to sacrifice the very constitutional bases for our actions that we are anxious to preserve.

When I read commentators blithely calling the president's targeted killing policy unlawful or unconstitutional or even a war crime without the slightest attention to legal argument and its forms, I despair because I know that many in our public won't be able to tell the difference. Law is not a science, but it is not just a matter of opinion either. Without some grounding in American legal precedents, rationales, and practices, how can a member of the public tell whether what the public is being told about the content of the law is true or false?

This book provides the answer: by reading an example of scrupulous, impeccable legal analysis. It is a gift to our democracy, and there was rarely a time when such a gift was more needed.

Philip Bobbitt
Herbert Wechsler Professor of Federal Jurisprudence,
Columbia University
Distinguished Senior Lecturer,
University of Texas School of Law

ACKNOWLEDGMENTS

In writing this book I have incurred many debts. My wife, Sarah, spent many invaluable hours with me thinking about the questions and the possible answers. I first tested the core of my argument in great conversations with my daughter Sara over coffee on the deck, and through e-mail with Walter Dellinger. Five other good friends gave me the gift of what were truly extraordinary line-by-line critiques of the first draft: Philip Bobbitt, Kent Greenawalt, David Lange, Henry Monaghan, and Richard Pious. I am deeply grateful to each for his generosity and time. I have also benefited immeasurably from the criticisms and encouragement of Wells Bennett, Joseph Blocher, Tom Brennan, Jessica Bulman-Pozen, Suzanne Goldberg, Margaret Hu, Dawn Johnsen, John McCausland, Gillian Metzger, Ralf Michaels, Dave Pozen, Matt Waxman, Lloyd Weinreb, and Jim White. It is a great honor to have a foreword written by Philip Bobbitt, but this is only the most recent of the many ways he has given me unfailing support over the years. I had the opportunity to present my ideas at workshops early in the writing process at Vanderbilt and Minnesota Law Schools and, after I had finished a rough draft, at Columbia

Law School, and I am grateful to the faculties at those schools for their interest and their valuable comments and questions.

The library staffs at the Duke and Columbia Law Schools provided invaluable help in research. In particular, my liaison librarian at Duke, Marguerite Most, was (as always) tireless and insightful. Brittany Edwards-Franklin (Duke J.D. 2014) surveyed the blogosphere for me with a thoroughness I could not have matched. My assistants Victoria Zellefrow and Sara Kuperminc showed their usual patience and efficiency, especially in fixing the results of my all too frequent mishaps with technology. My daughter Zoe prudently avoided conversations about the book, inadvertently reminding me of the truth that the Republic exists to serve humanity and not the other way around.

The reader may wish to know that I served in the Office of Legal Counsel of the U.S. Department of Justice in 2011 and 2012. During that time, I was never involved in any conversations involving targeted killing or the use of drones, nor did I see any documents related to these matters: I have no information that is not publicly available. The views I express are my own, and do not always coincide with those of the Department.

For many years David Lange has been my closest faculty colleague, my intellectual companion and co-author, and my dear friend. David is a brilliant intellectual property lawyer, but his knowledge and interests are catholic, and he is not simply learned but also wise: I have learned more from him than I can say. David and his equally gifted wife, Terri, are Sara's godparents, and they have been, through all this time, the friends on whom I have always relied. Words are quite inadequate in this situation, but I hope that David will understand what I mean.

INTRODUCTION

On September 30, 2011, the executive branch of the U.S. government deliberately killed a citizen of the United States. The executive's national security and intelligence agencies had identified the citizen, Anwar al-Awlaki, as a senior leader of al-Qaeda in the Arabian Peninsula, also known as AQAP, an Islamic jihadist organization that the executive views as a wing or affiliate of the original al-Qaeda, the organization responsible for the events of September 11, 2001.[1] The killing did not take place on a battlefield or in a conventional military operation against opposing forces; it was carried out by a "drone strike," an attack by munitions delivered by one or more unmanned aircraft. The strike was aimed specifically—and only—at al-Awlaki, although several other people, including a second U.S. citizen, Samir Khan, were also killed. Later that day, President Barack H. Obama hailed the killing as "a major blow to al Qaeda's most active operational

1. On July 12, 2010, the Office of Foreign Assets Control in the Department of the Treasury "determined that Anwar Al-Aulaqi was a key leader in AQAP who had been, and continued to be, involved in recruiting, training, and preparing terrorists for attacks on U.S. targets." Al-Aulaqi v. Panetta, 35 F. Supp. 3d 56, 60 (D.D.C. 2014). I see no reason to doubt that this determination was correct.

affiliate" and "another significant milestone in the broader effort to defeat al Qaeda."[2]

The executive branch did not decide to kill al-Awlaki on the spur of the moment or in the midst of a specific emergency: Officials had been discussing the attack for well over a year, judging by the Justice Department memo dated July 16, 2010, that concluded the killing would be lawful. There was not even anything very secret about the possibility that the Obama administration might use lethal force against al-Awlaki. While the Obama administration would fight hard to keep its lawyers' advice about the lawfulness of the killing under wraps after al-Awlaki was dead, before the drone strike the fact that killing him was under consideration for many months was so well known that his father had time to seek, in vain, the protection of the federal judiciary prior to the attack.[3] Two weeks after al-Awlaki died, another drone strike killed his sixteen-year-old son, Abdulrahman al-Awlaki, while the youth was sitting in a café or, by another account, eating a meal outdoors with his cousins. The younger al-Awlaki, also a U.S. citizen, was not the intended target, and President Obama was reportedly "surprised and upset" by his death.[4]

2. Remarks by the President at the "Change of Office" Chairman of the Joint Chiefs of Staff Ceremony (Sept. 30, 2011), https://www.whitehouse.gov/the-press-office/2011/09/30/remarks-president-change-office-chairman-joint-chiefs-staff-ceremony.

3. *See* Al-Aulaqi v. Obama, 727 F. Supp. 2d 1 (D.D.C. 2010) (dismissing the suit on standing and, in part, political question grounds). A different judge rejected a subsequent damages action for the killings of the two al-Awlakis and Khan. *See* Al-Aulaqi v. Panetta, 35 F. Supp. 3d 56 (D.D.C. 2014) (dismissing the suit on various grounds).

4. *Compare Panetta*, 35 F. Supp. 3d at 60 (the October 14, 2011, drone strike killed at least seven people at or near a café) *with* JEREMY SCAHILL, DIRTY WARS: THE WORLD IS A BATTLEFIELD 507 (2014) (Abdulrahman and several cousins, also teenaged, killed while at an outdoor picnic). One of the many books about the war on terror that have already appeared, Scahill's is especially valuable because of its focus on Anwar al-Awlaki although the reader should bear in mind that Scahill is an

The killing of al-Awlaki was not the first U.S. drone strike in the post-9/11 war against international terror. The "targeted killing" of high-ranking members of al-Qaeda and its allies, by drone or commando-style operation, began as a subordinate part of the U.S. response to 9/11 during the presidency of George W. Bush, but by the fall of 2011 it was arguably the signature element of the Obama administration's conduct of the war. Before al-Awlaki's death, among those particularly interested in national security matters a lively debate over the wisdom, the morality, and the legality under international law of targeted killing was ongoing, but it seems fair to say that the American public was paying little attention beyond its diminishing interest in the details of the war on terror in general. Because al-Awlaki was a U.S. citizen, killing him transformed, at least temporarily, the public's hum-hum indifference. Politicians, pundits, and Americans previously unconcerned about drone strikes in far-off countries aimed at foreign terrorists, found the spectre of our government killing one of us frightening. The administration defended the killing as a major blow to "the most dangerous regional affiliate of al-Qa'ida": the elimination of AQAP's "chief of external operations" whose role included "direct personal involvement in the continued planning and execution of terrorist attacks against the U.S. homeland."[5] But for many critics, the targeted killing of Anwar al-Awlaki seemed more like a criminal assassination or extrajudicial execution than an act of war. Writing two days after, for example, Senator Rand Paul charged that the

investigative journalist with a highly critical perspective on the U.S. government's actions. On the president's reaction to the death of the younger al-Awlaki, *see, e.g.*, Obama "Surprised," "Upset" when Anwar al-Awlaki's Teenage Son Was Killed by U.S. Drone Strike (Apr. 23, 2013), http://www.huffingtonpost.com/2013/04/23/obama-anwar-al-awlaki-son_n_3141688.html.

5. Letter of Eric H. Holder, Jr., to Sen. Patrick J. Leahy (May 22, 2013), http://www.documentcloud.org/documents/703181-ag-letter-5-22-13.html.

killing showed that "President Obama apparently believes he is not bound by the Constitution or the rule of law. . . . Under our Constitution, American citizens, even those living abroad, must be charged with a crime before being sentenced."[6] Killing al-Awlaki may have advanced the government's efforts to defeat al-Qaeda, but whether that is true or not, the drone strike that ended his life opened up a new and quite different conflict, over the consistency of those efforts with the Constitution the government's officers are sworn to uphold.

This book proposes answers to the constitutional questions raised by the targeted killing of al-Awlaki.[7] Anwar al-Awlaki was killed by a drone rather than by a sniper or cruise missile or in the course of a raid by U.S. special operations personnel. I shall focus on killing by drone strike, the most novel and most controversial method of targeted killing, but most of my argument will apply equally to these older and more conventional means of using lethal force. Because the killing of al-Awlaki was not an isolated action, but part of a broader policy involving the use of drone strikes to eliminate key members of al-Qaeda and associated organizations, I will examine the killing by setting it within the broad legal framework that the Constitution of the United States creates to govern all actions by the government of the United States. Because the constitutionality of the killing is a different question from the issues of whether it was wise or moral or legal under international law, this book will address

6. Rand Paul, An Unconstitutional Killing: Obama's Killing of Awlaki Violates American Principles (Oct. 2, 2011), http://www.nydailynews.com/opinion/unconstitutional-killing-obama-killing-awlaki-violates-american-principles-article-1.961079.

7. As Chapters Two and Three show, serious constitutional issues lie underneath the question of whether the executive branch is acting consistently with the relevant acts of Congress, and so Chapters Four and Five will discuss questions of statutory construction as well.

the issues of constitutional law rather than the other concerns al-Awlaki's death raises, although as we shall see one consequence of doing our constitutional homework well is to underline the importance of asking these other kinds of questions. Focusing on constitutionality is not, therefore, an implicit judgment about significance: I ask the reader to limit his or her attention to issues of constitutional law, for the most part, not because those are the only or even the most important questions raised by the drone strike that killed al-Awlaki, but in order that we may see more clearly one piece of the puzzle, one dimension of what the U.S. government did—and is doing—in the war against international terror. Getting straight the constitutional issues is essential to reaching a judgment on the government's action, not the entirety of such a judgment.

This book makes two presuppositions that I want to make clear at the outset. The first presupposition concerns the book's subject, the U.S. government's killing of Anwar al-Awlaki, and many others before and after September 30, 2011. Much of the public debate seems to assume that there is something radically new about "targeted killing," but to think so is to begin thinking about the killing of al-Awlaki by making a radical mistake. Targeted killing—the deliberate killing of a specific enemy—is as old as warfare itself. Face-to-face personal combat is, ordinarily and almost by definition, directed against particular foes, and in that sense most military killing in the premodern world was "targeted." The bow and arrow permitted a different kind of targeting, the killing of an enemy because of the tactical or strategic advantages of killing that specific individual rather than because he is trying to kill the bowman. In Book VIII of the *Iliad*, when the Greek generalissimo Agamemnon praises him for "laying waste from the strong bow the Trojan battalions," the archer Teukros explains that he has been trying to kill the great Trojan hero Hektor: "since we began driving the Trojans back upon Ilion . . . I have been lurking here with my bow, to strike down/ fighters . . . yet still I am not able to hit this mad dog."[8] The introduction of firearms gradually reduced to minor importance the role of hand-to-hand combat in conventional war, while providing new means of targeting particular enemy soldiers to achieve particular military goals. The sniper or marksman is Teukros's obvious descendant, but aircraft, guided munitions, cruise missiles, and of course the drone all provide the means to attack enemies identified by role if not by name.

In one sense, then, nothing has changed radically in the twenty-first century. Even under the conditions of modern, high-intensity, state-to-state war with its extremely destructive weaponry typically directed against numerous targets at once, the deliberate elimination of specific combatants or small groups of combatants, from the general controlling the enemy forces to the private manning a crucial guardpost, is a potentially important tool of war. Contemporary military thought emphasizes the importance of degrading a conventional enemy's command and control systems, systems that are at one and the same time extremely sophisticated and vulnerable to disruption in a variety of ways, including the physical destruction of the infrastructure and the killing of the specific personnel who make it work.[9] The "law of war," the international legal norms governing warfare often now called the law of armed conflict or international humanitarian law (IHL), permits "targeting" in this general

8. ILIAD VIII.295–99. The translation is Richmond Lattimore's.

9. *See, e.g.*, MILAN N. VEGO, JOINT OPERATIONAL WARFARE: THEORY AND PRACTICE VIII-56 (2007): "In terms of the effects on the enemy command and control, the most important elements of [command and control warfare] are physical destruction and electronic warfare."

sense, and indeed requires it since the indiscriminate use of force is prohibited.

What is now called asymmetric military conflict, between a traditional nation-state and more amorphous organizations or movements—the paradigm, of course, is the "war on terror"—puts an even higher premium on the ability of a conventional military such as that of the United States to attack specific, legitimate enemy combatants. The enemy usually controls no territory in a traditional sense, wears no uniforms and makes no sharp distinction between military personnel and civilian adherents, uses civilian populations as camouflage and cover for its activities. In such circumstances, IHL may *require* a law-abiding belligerent to pinpoint the objects of lethal force with extraordinary care in order to avoid violating the principles of distinction (which identifies who may and may not be deliberately killed) and proportionality (which forbids attacks on legitimate military targets when, among other considerations, the potential loss of civilian lives would be excessive relative to the military purpose of the attack).

There is, then, nothing novel about the general idea of targeted killing in warfare; in Western culture there has always been an undercurrent of worry as well about targeting an enemy when the killing takes place outside the context of personal combat. Agamemnon may ruefully remark, of the archer who wounded his brother Menelaos, that the deed is "glory to him, but to us a sorrow," but the *Iliad* also reflects a warriors' code of honor according to which killing from a safe distance could be the mark of a coward. In Book XI, for example, Alexandros (Paris, Helen's Trojan husband) puts three Greek stalwarts out of action by archery fire—a signal if temporary military advantage for the Trojans—but the greatest warrior among those he injures, Diomedes, denies that there is any honor in the success: "You archer, foul fighter . . . If you were to make trial of me in strong combat with weapons/your bow would do you no good at all, nor your close-showered arrows . . . this is the blank weapon of a useless man, no fighter."[10] Targeting a great figure for death from afar is associated at times with a different form of dishonorable behavior, deceit and the violation of one's oaths and obligations: The bowman who shot Menelaos broke a truce in doing so, and Agamemnon prophesies that Zeus "himself shall shake the gloom of his aegis over all of them/in anger for this deception."

Both Homeric concerns, that some killing in war violates shared expectations about what is chivalrous or honorable—"that's just not done"—and that some forms of faithlessness and deceit are unacceptable even if directed against an enemy one may otherwise kill, reappear at the dawn of the modern law of war. In Hugo Grotius's widely influential 1625 treatise *De Jure Belli ac Pacis* (*On the Law of War and Peace*), Grotius argued that killing by means of poison or a poisoned weapon, and in some circumstances by assassination, is contrary to the agreed-upon law of civilized nations.

Writing about the use of poison, Grotius acknowledged the old idea that personal combat is a more fitting mode of warfare than other means of killing, while rejecting the idea that poisoning an enemy is immoral per se:

> If you take account only of the law of nature, in case it is permissible to kill a person, it makes no difference whether you kill him by the sword or by poison. By the law of nature, I repeat, for it is indeed more noble to kill in such a way that he who is killed may have a chance to defend himself; but this is not an obligation due to one [who it is permissible to kill].

10. Iliad XI.385–90.

Nevertheless, Grotius continued, "the law of nations—if not of all nations, certainly of those of the better sort—has been that it is not permissible to kill an enemy by poison." The true rationale for this rule, in Grotius's view, lies not in chivalrous ideals but in

> a consideration of the common advantage, in order that the dangers of war, which had begun to be frequent, might not be too widely extended. And it is easy to believe that this agreement originated with kings, whose lives are better defended by arms than those of other men, but are less safe from poison, unless they are protected by some respect for law and by fear of disgrace.[11]

Grotius's discussion of assassination is similar: Those who "are held by no bond of good faith" may kill an enemy otherwise permissible to kill, "in any place whatsoever; and it does not matter how many there are who do the deed, or who suffer." Targeting a specific enemy, and attacking him "in his bedchamber" or by a sneak attack in "the hostile camp of the enemy . . . in the tent of the king" does not affect the lawfulness of the killing.

> But a different point of view must be adopted in regard to those assassins who act treacherously. Not only do they themselves act in a manner inconsistent with the law of nations, but this holds true of those who employ their services. . . . The reason . . . is the same that we advanced above with regard to the use of poison. It has in view the purpose to

11. Grotius, On the Law of War and Peace IV.15, translation in 1 The Law of War 37 (Leon Friedman ed., 1972).

> prevent the dangers to persons of particular eminence from becoming excessive.[12]

In many ways that I shall not examine, contemporary international humanitarian law has moved far beyond the thought of Grotius and other early international lawyers, but Grotius's conclusions that the law of war permits attack, including surprise attack, on specific enemies but prohibits the use of poison and assassination (in the sense of killing someone through treachery or breach of faith) are the modern norm as well. Grotius's prohibitions have been formally adopted and are enshrined in international conventions and in the International Committee of the Red Cross's widely accepted restatement of customary IHL.[13] In contrast, the formal sources of IHL recognize the legality of targeted killing in the general sense by the oblique means of not prohibiting it, but their logical import is clear; as a distinguished commentator has put it:

> All soldiers who are not injured or captured are presumed to be "seeking to kill," and therefore the deliberate targeting of all enemy combatants—regardless of their role on the battlefield, whether they are off duty, asleep in the barracks, bathing in the lake, or retreating—is permissible. There is no duty to warn enemy combatants prior to an attack, to try to capture enemy combatants as POWs instead of killing them (a duty that existed under the medieval codes of knightly conduct in respect to other knights), to injure them instead

12. *Id.* IV.18, in 1 THE LAW OF WAR, at 39–40.

13. *See, e.g.*, International Committee of the Red Cross (ICRC), *Customary IHL*, Rules 65 (assassination) and 72 (poison and poisoned weapons).

of killing them, or in any other way minimize combatant casualties.[14]

In particular, a decision to target "an officer of such high regard that his death would constitute a significant degradation of enemy warfighting capability [is] a perfectly legal and acceptable action in the conduct of warfare: provided such actions are taken openly and not through the use of what the Geneva Conventions describe as 'perfidy.'"[15]

The use of unmanned aircraft to attack specific individuals believed to be international terrorists is nothing that Grotius—or Homer!—ever imagined, but the current debate over the U.S. government's use of drones reflects fundamental issues that Grotius addressed. Warfare by definition and of necessity involves the use of lethal force against human life, but not even the Homeric hero thought that war licenses all killing, at any time and by any means, as long as the target is thought an enemy. If we are to think clearly about the killing of Anwar al-Awlaki, we need to recognize that there is a very real sense in which that killing simply displays the truth about warfare, that it *is* about killing, and about choosing those whom we decide to kill. Putting a name and a face, and a nationality, to the person the U.S. government targeted for lethal force on September 30, 2011, is disturbing in a way that reports about "enemy casualties" are not, but it serves no purpose to pretend that there is an absolute qualitative gap between the traditional means of fighting a war and the deliberate selection of a particular individual to eliminate: There is not.

14. Gabriella Blum, *The Dispensable Lives of Soldiers*, 2 J. Legal Analysis 115, 126 (2010). Professor Blum's article does not endorse the current IHL rules; she argues instead for enhanced limitations on the vulnerability of combatants to attack with lethal force.

15. Thomas B. Hunter, Targeted Killing: Self-Defense, Preemption, and the War on Terrorism 18 (2009).

At the same time, if we are to think clearly about the killing of al-Awlaki, we cannot escape in the end questions that echo Homer's and Grotius's concerns over what counts as the proper use of lethal force, over what is war and what are its limits. When we raise those concerns about the killing of al-Awlaki and others by drone, moreover, we confront a norm that neither the ancient poets nor the early modern international lawyers knew: the U.S. Constitution. In this Republic, the questions of whom the United States may kill as an act of war, and how and when and where, are issues of constitutional law in the first instance, however important other norms may be.

That brings us to the book's second presupposition, which concerns the nature of U.S. constitutional law. Because the Constitution is a written document, it is easy enough to assume that constitutional law is simply a matter of matching the meaning of the document to the question one is addressing. That conception of constitutional law isn't wrong, but American history suggests that it very seriously oversimplifies the process by which the Constitution actually functions as the framework of government. Felix Frankfurter, at the time a professor but later, and for many years, a Justice of the U.S. Supreme Court, once wrote about the great Justice Oliver Wendell Holmes, Jr., that for Holmes "the Constitution is not a document for fastidious dialectics but the means of ordering the life of a people.... Whether the Constitution is treated primarily as a text for interpretation or as an instrument of government may make all the difference in the world."[16] Frankfurter's point, to be sure, was not to depreciate the significance of the constitutional text but to make it clear how one is to approach it. Constitutional law reasoning is

16. Felix Frankfurter, *Mr. Justice Holmes and the Constitution*, 41 Harv. L. Rev. 121, 131 (1927).

a form of legal reasoning in which there is an authoritative text. But constitutional thought, in the American tradition, cannot be reduced to algorithmic form or limited to the value-free exercise of professional judgment: Constitutional judgments are value-laden. As Justice Byron White once wrote,

> The Constitution is not a deed setting forth the precise metes and bounds of its subject matter; rather, it is a document announcing fundamental principles in value-laden terms that leave ample scope for the exercise of normative judgment by those charged with interpreting and applying it.[17]

Responsible constitutional reasoning cannot proceed as if constitutional decision could be assimilated in scope and method to the formalisms employed in real estate law. The Constitution is both the document (or rather, including the amendments, the set of documents) *and* the "fundamental principles" of government that the document announces.

Understanding constitutional law in this manner, which I believe is the traditional American understanding, has important implications for the project of this book, to analyze the constitutional validity of the killing of al-Awlaki and more generally of the U.S. government's current targeted killing policy. Constitutional thought is not limited to the application of a lawyer's formal techniques of textual interpretation and the identification of preexisting understandings, although those formal modes of reasoning are essential. The very language of the text to be interpreted is "value-laden," and responsible constitutional

17. Thornburgh v. Am. Coll. of Obstetricians & Gynecologists, 476 U.S. 747, 789 (1986) (dissenting opinion), *overruled by* Planned Parenthood v. Casey, 505 U.S. 833 (1992).

decisions demand "the exercise of normative judgment" about the meaning of principles and values, not just the metaphorical "metes and bounds" of the Republic's government. Constitutional arguments are arguments over issues of *law*, but *constitutional* law is distinct from other areas of law because it deals with "the basic charter of our society," which "set[s] out in spare but meaningful terms the principles of government," and demands what White described as "normative judgment" about the content and application of those principles.[18]

The balance between the exercise of such judgment and the exercise of formal technique varies from question to question. Many conceivable constitutional issues are governed by a specific constitutional provision or have been the subject of repeated discussion and decision. A good example of this sort of issue are the questions that arise under the Emoluments Clause of Article I section 9 of the Constitution, which provides that "no Person holding any Office of Profit or Trust under [the United States], shall, without the Consent of the Congress, accept of any present, Emolument, Office, or Title, of any kind whatever, from any King, Prince, or foreign State." With some frequency, foreign governments and governmentally related entities offer benefits and honors to persons employed by, or with some other connection to, the federal government. Does the Emoluments Clause, for instance, prohibit a scientist serving as an unpaid member of an advisory committee to a federal agency from accepting a visiting professorship at a public university in a foreign country? The question is constitutional, and in its own, rather minor-key way implicates important issues about loyalty and allegiance, but in answering it one can and should draw on the rich body of

18. The first quotation is from Justice John Marshall Harlan's famous dissent in *Poe v. Ullman*, the first contraception case to reach the Supreme Court. *See* Poe v. Ullman, 367 U.S. 497, 540 (1961) (Harlan, J., dissenting).

legal reasoning and decision that has developed to answer such questions.[19]

The constitutional issues raised by the killing of Anwar al-Awlaki, in contrast, go straight to heart of what constitutional government in this Republic means: where in government authority exists to engage in military action, the conditions under which government may kill human beings, the relevance of U.S. citizenship to such a decision. There are, to be sure, several provisions in the text of the Constitution that seem relevant—the clauses granting Congress the power to declare war and making the president commander in chief come to mind—but there is surprisingly little authoritative discussion about how those potentially conflicting provisions, and others such as the Fifth Amendment's Due Process Clause, apply to specific issues. In addressing the constitutional issues raised by the killing of al-Awlaki, technical professional arguments will play a lesser role, and normative judgments about the fundamental principles of American government must lie at the center of any convincing answers.

* * *

Our inquiry into the constitutionality of killing Anwar al-Awlaki begins by placing the drone strike into its historical context. "Historical" here does not mean the general history of the American "war on terror" that began on September 11, 2001. That history is an urgent and frequently tragic story that is still

19. As a functional matter, it is the executive rather than the judicial branch that bears primary responsibility for adherence to the Emoluments Clause, and so (unusually in constitutional law) executive branch legal opinions rather than judicial decisions are the primary source of legal authority. *See, e.g., Application of the Emoluments Clause to a Member of the Federal Bureau of Investigation Director's Advisory Board*, 2007 WL 2284482 (O.L.C. 2007). The Office of Legal Counsel (OLC) assists the attorney general in providing authoritative legal advice to executive branch officials. The reader may breathe a sigh of relief that we shall not explore the Emoluments Clause, or the constitutionality of visiting professorships abroad, further.

unfolding, and even apart from its unfinished nature it is very difficult to describe with any confidence what has happened because the major actors, including the U.S. government, are so secretive, and the reports of events regularly conflict. Most of its details are not, however, crucial to examining the constitutional issues presented by the al-Awlaki killing. With the exception of a brief section in Chapter Five, for our purposes we shall not need to consider or resolve any of the many hotly contested claims. I assume throughout what is reasonably clear and in any event hardly contested even by critics of the Obama administration: Anwar al-Awlaki was a very senior figure in AQAP whose role and activities made him the equivalent in al-Qaeda of a general in a conventional military. He was, in short, an enemy combatant rather than a civilian, and the U.S. government was legally entitled to treat him as such.[20] The history that is relevant to our argument and that is the main subject of Chapter One is the *constitutional* history of the federal government's actions against international terrorists, which at the most basic level is the narrative of changing positions and actions of the two political branches of the government with respect to terrorism.

Chapter One accordingly distinguishes four stages or phases in the war on terror. In Phase One, the period before 9/11, it is inaccurate to speak of a "war" on terror: Congress had not

20. As is true about so much concerning the war on terror, a straightforward statement oversimplifies complex and contested reality. There are serious questions about which body of international law, IHL or international human rights law, properly governs the U.S. conflict with al-Qaeda and its allies, and (relatedly) about how someone like al-Awlaki should be categorized for international law purposes. The U.S. government position is that IHL is the governing international law and that military figures in al-Qaeda and its regional affiliates are unlawful enemy combatants. For reasons that will become clear in Chapters Three and Four, given what the executive believed was factually true about his activities, as a matter of American constitutional law al-Awlaki was a member of an enemy's armed forces and lawfully treated as such.

authorized military action and the executive did not claim that as a legal matter there was a war in the constitutional sense. Phase Two began when the attacks of 9/11 created an undeniable national emergency to which the president was constitutionally authorized to respond with military measures that rose to the level of making war, and ended a few days later when Congress approved the use of military force against the perpetrators of 9/11 by an Authorization for the Use of Military Force (the 2001 AUMF). The new constitutional Phase Three initiated by Congress's action complicates the analysis, since (as I will argue) the AUMF limits as a general matter the executive's use of military force in what we now can label the war on terror. It was during Phase Three, which is still going on as of the time of writing, that the U.S. government killed al-Awlaki. Finally, as the circumstances and enemies addressed by Congress in 2001 recede into the past, and the executive's focus in the war on terror shifts to new enemies and new problems, we begin to emerge into a constitutional Phase Four, with uncertain consequences for the executive's authority to engage in targeted killing or, indeed, other military measures.

In Chapters Two and Three, we turn to the conceptual context in which the constitutionality of killing al-Awlaki should be considered. It is impossible to resolve a novel or difficult constitutional issue if one approaches it either as a narrowly professional question of technical law or as a free-form exercise in draping constitutional rhetoric over a conclusion reached on extralegal, extraconstitutional grounds. Fortunately, the American constitutional tradition has developed a mode of constitutional reason that is intellectually expansive enough to deal with grave and unprecedented issues, and to do so in an intellectually rigorous manner. Chapter Two discusses how to "think constitutionally" within this tradition: Sound constitutional conclusions respect the authority of the constitutional text but read that text in the light of the Constitution's great purposes.

Chapter Three outlines the doctrinal framework by which constitutional law structures issues about war. The federal government's war powers are plenary—they should be read to empower the government to wage war successfully—but they are divided among the two political branches, and except for the president's exclusive function as supreme commander, congressional decisions control presidential discretion in the event of a conflict. The war powers, like all federal authority, are also limited by applicable constitutional provisions that prohibit specific modes of exercising the government's powers, but Chapter Three shows that as a matter of principle and precedent most such provisions (including those in the Bill of Rights) do not apply to the exercise of the powers to wage war.

Chapter Four attempts to think constitutionally about the targeted killing of al-Awlaki. Because the killing took place in what Chapter Two refers to as Phase Three of the war on terror, the 2001 AUMF was the relevant source of legal authority for the executive's decision to target al-Awlaki, and the killing was in fact within the authority Congress conveyed to the executive. The widely debated question whether the killing violated the Due Process Clause of the Fifth Amendment rests on a mistake: that clause does not protect enemy combatants, including U.S. citizens, from lethal war measures. If the clause does apply, as many people including the Obama administration's lawyers believe, the killing did indeed deny al-Awlaki due process and was unconstitutional.

In Chapter Five, the book turns to the future and to three very difficult questions. First, is the 2001 AUMF a grant of power to fight terrorism in perpetuity until Congress amends or repeals it? The arguments on balance favor a negative answer: At some point, as time passes and the executive itself believes and asserts with increasing confidence that it has substantially defeated the original enemies, it will no longer be an act of constitutional good faith for the executive to rely on the AUMF, and this new Phase Four will come to resemble Phase One, with the president relegated to whatever authority he or she has to take military action without legislative authorization. Second, would it be constitutional for Congress to prohibit the executive from employed targeted killing by drone as a means of conducting lawful hostilities in the war on terror? This question poses against one another core constitutional principles in the area of war powers: Congress's powers to circumscribe the president's use of force and to enact rules governing the conduct of military personnel versus the president's power as final military decision maker on issues of tactics. It is impossible to be sure about the better answer, but I conclude with hesitation that a congressional ban on drone strikes would be valid. Finally, are there truly no outer constitutional limits to the scope of the combined war powers of the political branches? The easier answer is no: The Constitution assumes that Congress and the president will attempt to exercise strategic wisdom, refrain from barbaric actions, and at least generally follow the norms of international law, but does not put significant legal limits on their discretion. I consider the possibility that the war powers may not authorize targeted killing by drone in two situations, where the killing of a specific individual amounts to assassination and where drone strikes become so indiscriminate that they become themselves a mode of terrorism. The argument against an assassination limitation on the war powers is unconvincing, at least where Congress has authorized hostilities, but there is reason to take seriously the argument that the Constitution does not empower the federal government to take military actions that contravene the very basis of the American political community.

We must ask no more—and no less—of constitutional law than constitutional law can provide. This book is a sustained effort to think through the constitutional legitimacy of the U.S. government's action in killing Anwar al-Awlaki. In the conclusion, I raise the question of relevance: How and why is it important to get our constitutional thinking right? I propose a three-part answer. First, sound constitutional analysis is conducive to good governmental decision making. Second, good faith in constitutional reasoning is essential to ensuring the legitimacy of governmental actions governed by the Constitution. Sound constitutional thinking, finally, enables us to see the limits of constitutional inquiry and to ask questions about the wisdom and the morality of governmental action that have no constitutional answers. Such considerations are always relevant to the exercise of political power. When the government of the United States decides on our behalf to kill, the ultimate goal of American constitutionalism is to demand more of the government than mere legality.

Chapter 1

The Constitutional History of the War on Terror

Before the twenty-first century there was no distinctively American perspective on targeted killing because there was no distinctively American experience on the matter. The main current in American military thought, as with Western thought generally, has accepted the deliberate targeting of specific enemy combatants as simply one of the many legitimate ways to employ lethal force in war. In the Revolutionary War, none other than George Washington, that American paragon of military probity, approved the formation of units of marksmen expert in using the rifle, a weapon both longer-range and more accurate than the ordinary smoothbore musket of the era. Washington's "Provisional Rifle Corps," commanded by Colonel Daniel Morgan, gained a reputation for the effective use of sharpshooting to harass British musketeers with relative impunity, and in particular for picking off enemy officers: Morgan instructed his riflemen to "'shoot for the epaulettes' [because] it was an item of uniform worn only by officers." During the battle of Bemis Heights during the Saratoga campaign in 1777, one of Morgan's men, Timothy Murphy, perched in a tree in order to target and kill the mounted officer vigorously directing the British troops at a crucial point in the action. One of Brigadier Simon Fraser's subordinates thought his death by Murphy's targeted killing enabled the Americans "to turn the fate of the day."[1]

Later wars of the nineteenth and twentieth centuries were no different. The Lieber Code, the U.S. Army's 1863 codification of the international law of war (and the first modern restatement of that law), generally permitted the killing of enemy combatants, and rejected any extralegal "restriction of the modes adopted to injure the enemy." "Outside of the rules Lieber laid down, war could, and should, be prosecuted with the utmost ferocity."[2] Both the Union and the Confederate armies acted on the principle and recruited sharpshooters. "Sniping was a fundamental reality of Civil War military life, and rifles made marksmen accurate up to a distance of almost half a mile. Other technological innovations, the telescopic sight and the breech-loading rifle, further enhanced the sharpshooter's lethality." At the same time, Americans have not infrequently expressed the age-old worry that there is something dishonorable or particularly inhumane about targeted killing from a distance. For many soldiers in the Civil War, "[t]he cool calculation, the purposefulness, and the asymmetry of risk involved in sharpshooting rendered it even

1. Martin Pegler, Out of Nowhere: The History of the Military Sniper 39 (2004); Michael O. Logusz, With Musket and Tomahawk: The Saratoga Campaign and the Wilderness War of 1777, at 276 (2010).

2. *Instructions for the Government of the Armies of the United States in the Field*, Art. 30 (General Orders No. 100) (promulgated Apr. 24, 1863), in 1 The Law of War 164 (Leon Friedman ed., 1972); Paul Finkelman, *Francis Lieber and the Modern Law of War*, 80 U. Chi. L. Rev. 2071, 2097 (2013) (discussing Article 30). The Lieber Code followed the tradition going back to Grotius of prohibiting the use of poison or perfidy. *See also Instructions*, Art. 16 ("Military necessity . . . does not admit of the use of poison in any way"); Art. 101 ("the common law of war allows even capital punishment for clandestine or treacherous attempts to injure an enemy"), in 1 The Law of War, at 161, 177. Article 148 stands alone in a section entitled "Assassination," but its text refers only to "murder" committed pursuant to a proclamation that an enemy individual is an outlaw, and to "offers of rewards for the assassination of enemies." *Id.* at 184.

more threatening to basic principles of humanity than the frenzied excesses of heated battle." The fact that enemy marksman sometimes struck when their targets were off the battlefield and engaged in peaceful activities offended many Americans' sense of moral propriety, even though both the Union and the Confederacy employed sharpshooters. "Soldiers in camp wanted to think themselves off duty as target as well as killers, and they found the intentionality and personalism involved in picking out and picking off a single man highly disturbing."[3] Targeting a specific enemy combatant is not a product of this century's war on international terrorism; nor is concern over its legitimacy, especially when targeting takes place outside the context of immediate combat.

The most famous targeted killing carried out by the United States before 9/11 was undoubtedly the successful attempt during World War II to shoot down Japanese Admiral Isoroku Yamamoto. Originally a vigorous opponent of war with the United States (a war he did not think Japan could win), after the decision for war was made, Yamamoto oversaw the planning and execution of the attack on Pearl Harbor and commanded the Japanese Combined Fleet in its aggressive and largely successful campaigns in the following year and a half. Senior officers in the U.S. Navy held him in high respect and believed that "the Japanese navy had no comparable leader of strategic vision, stature and daring." At the same time, for most Americans, in and out of the armed forces, Yamamoto "was the hated face of Japanese war machine." When American military intelligence intercepted and decoded Yamamoto's flight plans for an inspection tour of Japanese advance positions near Guadalcanal in

3. Drew Gilpin Faust, This Republic of Suffering: Death and the American Civil War 42 (2009).

April 1943, the information presented the Navy with a chance of eliminating a feared enemy leader, inflicting a serious blow to Japanese morale, *and* taking revenge for Pearl Harbor.[4] The decision whether to make the attempt had to be made quickly—the radio interception occurred on April 14 and the most vulnerable leg of Yamamoto's trip was to take place on April 18.

Despite the obvious potential benefits, not everyone involved in the highly confidential decision-making process agreed that Yamamoto should be targeted. Acting on the information might reveal to the Japanese military the fact that U.S. cryptographers could read some of its supposedly unbreakable codes. The resulting Japanese switch to different and secure codes, some argued, "would be a military intelligence disaster far outweighing any possible gains accruing from the admiral's death." Others were concerned that killing Yamamoto would be a serious strategic and political mistake because of his realism about Japan's likely defeat: He was the only senior figure holding that view with the stature to challenge the unrealistic and hardline views on the war dominant in the Japanese war cabinet. From the perspective of these skeptics, the desire to kill Yamamoto was driven by emotion rather than cool strategic calculation, a "macho-military blunder, motivated solely by pique."[5] Some opponents of targeting Yamamoto, finally, may simply have found the action unchivalrous or morally objectionable.[6] Ultimately, and

4. Donald A. Davis, Lightning Strike 228–29 (2005).

5. *Id.* at 289–90.

6. It is possible that some involved in the decision to kill Admiral Yamamoto shared the views of John Paul Stevens, one of the U.S. Navy cryptographers at the time and later a distinguished member of the Supreme Court. In an article based on interviews with Justice Stevens, Professor Jeffrey Rosen wrote:

> Stevens told me he was troubled by the fact that Yamamoto, a highly intelligent officer who had lived in the United States and become friends with American officers, was shot down with so little apparent deliberation or

for whatever combination of reasons and motivations, Admiral Chester W. Nimitz ordered his subordinates to make the attempt, which was successful; it seems that President Franklin D. Roosevelt approved the killing without quite issuing a direct order to carry it out.[7]

The targeted killing of Admiral Yamamoto was both similar to, and very different from, the targeted killing of Anwar al-Awlaki. We shall discuss below in more detail the Obama administration's legal analysis of the al-Awlaki killing, but at its heart, that analysis treated killing al-Awlaki as legally indistinguishable from killing Yamamoto, who, by the way, was not explicitly mentioned. Each target was a high-ranking officer in the military forces of an enemy with which the United States was at war. The fact that manned fighter planes shot the admiral out of the sky while an unmanned aircraft killed al-Awlaki on the ground was of no more legal significance than the fact that the fighters used cannon and machine guns, while the drone fired missiles.

However, even if that conclusion is correct, there are obvious differences that must be addressed. No serious question existed about the lawfulness of killing the admiral. There was no doubt that Yamamoto was an officer serving in the armed forces of a nation against which Congress had specifically declared war; as a consequence, the constitutionality of killing him was beyond

> humanitarian consideration. The experience, he said, raised questions in his mind about the fairness of the death penalty. "I was on the desk, on watch, when I got word that they had shot down Yamamoto in the Solomon Islands, and I remember thinking: This is a particular individual they went out to intercept," he said. "There is a very different notion when you're thinking about killing an individual, as opposed to killing a soldier in the line of fire."

Jeffrey Rosen, The Dissenter, Justice John Paul Stevens, (Sept. 23, 2007), at http://www.nytimes.com/2007/09/23/magazine/23stevens-t.html?pagewanted=all.

7. *See, e.g.*, THOMAS B. HUNTER, TARGETED KILLING: SELF-DEFENSE, PREEMPTION, AND THE WAR ON TERRORISM 18 n.30 (2009) (noting the existence of debate over Roosevelt's involvement in the decision).

dispute. Under international humanitarian law (IHL: the law of war), furthermore, as a uniformed combatant Yamamoto was unquestionably a lawful target. At the time he was killed, Yamamoto was in fact engaged in carrying out his duties as such, although this latter fact is not a requirement of international humanitarian law when the object of lethal force is a lawful enemy combatant. The U.S. fighters who shot him down were operating in the theater of war—in effect, as much on a metaphorical battlefield as Brigadier Fraser was on a literal one. There were no issues about the risk of potential, unintended civilian casualties outweighing the military value of the attack: Yamamoto was flying on an armed bomber escorted by fighters.

In contrast, al-Awlaki wore no military uniform, literal or metaphorical, and the organization that he "served," al-Qaeda in the Arabian Peninsula (AQAP), is not a recognized nation-state or, arguably, part of the congeries of enemies against which Congress authorized military action in the wake of 9/11. At the time he was killed, al-Awlaki was engaged in, on the face of it, peaceful activities in the territory of a state (Yemen) that was not within the recognized American theaters of war in Afghanistan and Iraq; it was an exercise of at least theoretically contestable judgment that the U.S. government was entitled under IHL to treat al-Awlaki as a combatant subject to lethal action under the circumstances. The drone strike that killed al-Awlaki killed several other people as well, "collateral damage" that the government was willing to cause without regard to their status, unlike the Japanese personnel who died along with Admiral Yamamoto, who were themselves combatants and valid targets. The question whether these or other differences between the two targeted killings puts the killing of al-Awlaki in constitutional question is, of course, the subject of this book.

One last observation about the Yamamoto "precedent" is relevant. The unimpeachable constitutionality of targeting

Admiral Yamamoto played no known role in the debate over whether to kill him, and rightly so: Whatever part President Roosevelt may have played, the debate about Yamamoto was over a military action, the relevant issues were military ones, and the actual decision maker was, it seems, Admiral Nimitz, in his role as a military commander. In the painfully drawn-out deliberations over targeting al-Awlaki, on the other hand, constitutional concerns, and legal ones more generally, seem to have played a far greater role. This may be the product, in part, of an overall increase generally in the role of legal discussion, or perhaps better of lawyers, in internal governmental processes between 1943 and 2011, but it also suggests that officials in the Obama administration felt a greater interest in determining the lawfulness (or at least the appearance of lawfulness) of targeted killing than their predecessors in other wars have done. Focusing on constitutional issues may of course stem from an admirable concern about the legitimacy of governmental action, but doing so may also divert attention from other issues that ought to concern the decision makers: Constitutionality is a prerequisite for a governmental action, but not a sufficient reason for taking that action.

* * *

The factual context of the killing of Anwar al-Awlaki is of course the war on terror and the U.S. government's use of targeted killing by drone as a means of war making. The constitutional context of targeted killing by drone, however, primarily concerns not the detailed events of the conflict but rather the changing legal relationship between the legislative and executive branches of the federal government over the matter. We can usefully distinguish four phases in the evolution of that relationship.

Phase One. The constitutional history of targeted killing by drone begins several years before 9/11, during the administration of President Bill Clinton, as the executive branch gradually came to realize that Osama bin Laden and al-Qaeda posed an extremely serious threat not just to U.S. interests but to U.S. lives as well. When President Bill Clinton issued an executive order in January 1995 invoking economic sanctions on Middle East "terrorist" organizations that were endangering U.S. efforts for peace in the region, al-Qaeda was not even on the list, but in August 1996 bin Laden issued a "Declaration of War Against the Americans Occupying the Land of the Two Holy Places." Any doubt that the declaration was serious disappeared with al-Qaeda's deadly bombing of the U.S. embassies in Kenya and Tanzania in August 1998, in response to which U.S. Navy warships launched cruise missile attacks on a suspected al-Qaeda operation in Sudan and several training camps in Afghanistan. The Afghanistan strikes were intended to kill bin Laden and other terrorist leaders and were therefore the first known U.S. attempt to use targeted killing as a military tool against al-Qaeda.[8]

The Clinton administration publicly defended the missile attacks as consistent with international law but did not disclose what it believed to be the president's domestic-law authority to carry out the attack.[9] Despite the executive's coyness about the

8. *See* National Commission on Terrorist Attacks Upon the United States, The 9/11 Commission Report 116–17 (2004); Micah Zenko, Between Threats and War: U.S. Discrete Military Operations in the Post-Cold War World 13 (2010).

9. *See* Letter from Bill Richardson, Permanent Representative of the United States to the U.N., to Danilo Turk, President, U.N. Security Council, U.N. Doc S/1998/780 (Aug. 20, 1998) ("the United States has acted pursuant to the right of self-defense confirmed by Article 51 of the United Nations Charter. The targets struck, and the timing and method of attack used, were carefully designed to minimize risks of collateral damage to civilians and to comply with international law, including the rules of necessity and proportionality."). Following the statutory procedures Congress has imposed on covert operations, President Clinton also took legal steps intended to authorize the CIA to begin clandestine and potentially lethal actions against al-Qaeda. *See, e.g.*, Steve Coll, Legal Disputes over Hunt Paralyzed Clinton's

action's legal rationale, however, we can infer with considerable confidence the advice its lawyers gave. No act of Congress authorized, in terms or by implication, the use of lethal military force in either Sudan or Afghanistan or against al-Qaeda, and so the only plausible source of legal authority was the U.S. Constitution itself. (Theoretically a treaty could provide legal authorization, but no treaty to which the United States is a party was relevant in Phase One, or indeed after.) No one doubts that the Constitution authorizes the executive to respond to an actual attack, and the argument could be made that the bombing of U.S. embassies was the constitutional equivalent of bombing sites on U.S. soil. However, the Clinton administration itself clearly disavowed justifying the missile attacks as a discrete response to a specific attack: The chairman of the Joint Chiefs of Staff bluntly said that "[t]his is not simply a response to some specific act, but a concerted effort to defend U.S. citizens and our interests around the globe against a very real and a very deadly terrorist threat," while other executive branch officials described the missile attacks as "the start of what one called 'a real war against terrorism,' emphasizing that 'this is not a one-shot deal here.'"[10]

Without a statutory basis, or the existence of a national emergency, the lawfulness of the 1998 missile attacks had to rest on a broader presidential authority to use military force without legislative approval. The scope and even the very existence of such authority is highly controversial as a legal matter even if unilateral presidential action is familiar: Views on the matter range from a flat denial that any such power exists (popular among scholars)

Aides (Feb. 22, 2004), http://www.washingtonpost.com/wp-dyn/articles/A59781-2004Feb21.html.

10. U.S. Strikes Terrorist-Linked Sites in Afghanistan, Factory in Sudan (Aug. 21, 1998), http://www.washingtonpost.com/wp-srv/inatl/longterm/eafricabombing/stories/strikes082198.htm.

to a claim of plenary executive power to engage in full-scale war. Like the lawyers of every administration since World War II, Clinton's Justice Department took the position that Article II of the Constitution vests the president with some power to act unilaterally. A 1994 legal opinion, for example, concluded that the president may order military action "without prior authorization by the Congress" if the "circumstances [including "the nature, scope, and duration" of the action] . . . show that the operation is not a 'war' within the meaning of the Declaration of War Clause."[11] With this as their legal premise, it is easy to see how executive branch lawyers could advise that a once-off cruise missile strike on a limited number of targets would not rise to the level of " 'war' within the meaning of the Declaration of War Clause." Public statements about the strikes being the first step in a "real war" could be set aside as hyperbole, and the legality of possible future uses of force left for future resolution when and if such military actions were planned—and the lawyers' advice sought.

Phase One in our history represents what has historically been the political and constitutional norm: Congress's power to declare war or otherwise approve military action was dormant, and any power the president had to use military force was limited to whatever actions the Constitution authorizes him to undertake unilaterally. However, the significance of Phase One for our inquiry, the constitutionality of targeted killing by drone strike, is not limited to providing background. As we shall see below, the constitutional situation at the time of al-Awlaki's death may be coming to a close, and if that situation, Phase Three, does end unmistakably, it is possible that the lawfulness of future drone strikes should be evaluated as the exercise of unilateral presidential authority in a new Phase One.

11. *Deployment of United States Armed Forces into Haiti*, 18 Op. O.L.C. 173, 175–76, 177–78 (1994).

Phase Two. From a constitutional perspective, the next important events were the devastating attacks of September 11, 2001. The events of 9/11 dramatically modified the constitutional uncertainty of Phase One: In Phase Two, there was no question about the lawfulness of employing military force against the perpetrators of the September 11 attack. It is in fact beyond serious question that the president may respond with military force to such an attack: In the War Powers Resolution of 1973, for example, a Congress intent on cabining presidential authority to use military force conceded that the president has the "constitutional power" as commander in chief "to introduce United States Armed Forces into hostilities" pursuant to "a national emergency created by attack upon the United States, its territories or possessions, or its armed forces."[12] On any reasonable view of constitutional authority, the loss of life and the scale of the physical destruction on 9/11 justified a vigorous military reaction—the problem at the time was knowing against whom to use force.

This does not mean that Phase Two posed no constitutional issues. Even the most cautious legal advisor should have advised President George W. Bush that there is substantial authority for the view that in the wake of an attack, the Commander in Chief Clause gives the president wide discretion to employ effective military means to protect the nation against further aggression even if those means would otherwise be well beyond the president's unilateral authority.[13] But as a strategic matter, 9/11 was

12. 50 U.S.C. § 1541(c)(3).

13. *See, e.g.*, Prize Cases, 67 U.S. 635, 668 (1862) ("If a war be made by invasion of a foreign nation, the President is not only authorized but bound to resist force by force. He does not initiate the war, but is bound to accept the challenge without waiting for any special legislative authority. And whether the hostile party be a foreign invader, or States organized in rebellion, it is none the less a war"); United States v. Smith, 27 F. Cas. 1192, 1230 (C.C.D.N.Y. 1806) (Paterson, Circuit Justice) ("If indeed, a foreign nation should invade the territories of the United States, it would . . . be not only lawful for the president to resist such invasion, but

not Pearl Harbor, an effort by a major military power to disable a central element in the U.S. armed forces defending U.S. possessions and interests throughout the Pacific. The Japanese attack on the Pacific Fleet thrust full-scale war upon the United States; after it, Congress's decision to declare war in the fullest sense was a foregone conclusion. The events of 9/11, grievous as they were, presented a national security threat of a different order, and even apart from the problem of identifying the enemy, there was no certainty that Congress would ultimately authorize an unrestrained military response—as indeed it didn't. As a constitutional matter, Phase Two imposed on the executive a twofold duty, to preserve the nation against further harm while making it possible for Congress to execute its duty of determining what long-term steps the government should take against the attackers.

Phase Three. President George W. Bush's legal advisors did not have long to worry about the scope of his constitutional authority in Phase Two, however, because on September 18, Congress enacted an Authorization for the Use of Military Force (AUMF) against the perpetrators. This AUMF specifically approved the use of "all necessary and appropriate force against those nations, organizations, or persons he determines planned, authorized, committed, or aided the terrorist attacks that occurred on September 11, 2001, or harbored such organizations or persons."[14] With Congress's arrival on the constitutional stage,

also to carry hostilities into the enemy's own country."). Professor Charles Lofgren's careful study of the debate over the lawfulness of the Korean War concluded that both the supporters and the critics of the Truman administration agreed that in a Phase Two situation, the president has unilateral authority "to counter an immediate, dangerous threat to American interests and security." CHARLES A. LOFGREN, GOVERNMENT FROM REFLECTION AND CHOICE 225 (1986).

14. Pub. L. 107–40, § 2(a). The September 2001 AUMF should not be confused with the Authorization for the Use of Military Force against Iraq enacted in October 2002.

the presumptive basis for military action against the enemy (soon identified as bin Laden's al-Qaeda and its allies) became the congressional enactment: As President Obama put it in 2013, "[u]nder domestic law, and international law, the United States is at war with al Qaeda, the Taliban, and their associated forces."[15] Unsurprisingly, therefore, the September 2001 AUMF has been the primary domestic law basis on which Obama administration, and with some exceptions the Bush administration before it, have relied in the war against terror, and we shall examine it below in some detail.[16]

The September 2001 AUMF unmistakably authorized the use of conventional military means for striking back at al-Qaeda, and the legality under U.S. law of the Bush administration's initial steps was clear. After the Taliban government of Afghanistan rejected American demands to turn over bin Laden, in September and October 2001, U.S. forces cooperated with anti-Taliban fighters in Afghanistan to topple that government and drive its leadership, and that of al-Qaeda, into exile in remote areas on the Afghan border with Pakistan or elsewhere. The failure of the United States and its allies to completely destroy the Taliban or capture bin Laden turned that invasion into a conventional, long-running counterinsurgency effort to establish

15. Remarks by the President at the National Defense University (May 23, 2013), https://www.whitehouse.gov/the-press-office/2013/05/23/remarks-president-national-defense-university.

16. Shortly after 9/11, an official in the Office of Legal Counsel (OLC) of the Justice Department signed a legal opinion that treated the AUMF as having no effect beyond providing a "welcome" but legally irrelevant congressional blessing on "the President's unilateral war powers" to respond to the attack. *The President's Constitutional Auth. to Conduct Military Operations Against Terrorists & Nations Supporting Them*, 2001 WL 34726560, at *17, *19 (O.L.C. Sept. 25, 2001). While no subsequent executive branch legal opinion has disputed the continuing existence of some unilateral Article II power to take military action against terrorists, the executive has not adhered to this reduction of the AUMF to a congressional vote of confidence.

a stable and secure—and friendly—Afghanistan. The U.S. invasion of Iraq in March 2003 in order to topple the Saddam Hussein regime turned into a second counterinsurgency war, and the two conflicts largely merged in American public perception, although their legal bases were entirely distinct: Operations in Iraq were lawful under the Authorization for the Use of Military Force in Iraq that Congress enacted in October 2002.[17]

In Afghanistan (and Iraq) American targeted killings by drone strike or otherwise in this period took place in recognized theaters of war. They were merely one tool in broader military operations conducted by conventional U.S. ground and air units, and on the face of it any legal questions they raised could be subsumed into debate over the broader conflicts, a debate in which there was little apparent reason for constitutional concern, as I have suggested. But at the same time, the United States became involved in a second kind of war, a gradually widening struggle to locate and destroy elements or allies of al-Qaeda in countries where there was no public U.S. military presence, and the primary U.S. goals were not those of classical counterinsurgency efforts. Targeting key members of an enemy dispersed among populaces that the United States could not and would not police generally for political reasons was from the first an important element on this side of the war against terror, although reliance on the drone only slowly developed: Ostensibly more traditional forms of military action such as the commando-style kill/capture mission were and remain important. The most famous of all targeted operations in the current conflict, after all, was the killing of bin Laden by U.S. Navy Seals on May 2, 2011.

The drone strike made its first appearance other than as part of the Afghan conflict in Yemen in 2002 in an attack that killed its

17. Pub. L. 107–243 (Oct. 16, 2002).

intended target; the attack also killed several others believed to be al-Qaeda terrorists, including a U.S. citizen. Despite this tactical success, the Bush administration did not rely heavily on drones and apparently launched no further drone strikes in Yemen after 2002. In late 2009, however, the Obama administration began targeted killings, generally with the Yemeni government's permission, by cruise missile and drone.[18] The United States began targeted killings by drone in Pakistan in 2004 after that country's government agreed to permit U.S. drone strikes in return for a significant degree of influence over who was targeted and where, but the use of drones in Pakistan increased dramatically under President Obama even as relations with the Pakistani government cooled.[19]

The killing of Anwar al-Awlaki was part of this turn to targeted killing by drone in the circumstances of Phase Three, in which the executive enjoys broad congressional authorization to use military force. The existence of the AUMF changes the nature of the constitutional inquiry. The first question becomes a matter of interpreting not the Constitution but the AUMF itself: What exactly does Congress's language authorize? Since Congress cannot authorize the president to violate the Constitution, constitutional analysis must also address the issue of whether executive action violates some constitutional limitation that constrains both branches—the provisions of the Bill of Rights, for example. Perhaps the most difficult question is the extent to which the president may invoke his unilateral Article II powers—whatever their scope—to act beyond or even against what Congress has legislated.

18. *See, e.g.*, Open Society Justice Initiative, Death by Drone: Civilian Harm Caused by U.S. Targeted Killings in Yemen 31–33 (2015).

19. *See, e.g.*, Ian G. R. Shaw, The Rise of the Predator Empire: Tracing the History of U.S. Drones (Oct. 28, 2012), https://understandingempire.wordpress.com/2012/10/28/essay-the-historical-rise-of-the-predator-drone-2/.

Congress has not repealed the September 2001 AUMF, and at the time of writing there appears to be little chance that it will. The statute does not contain an express sunset provision putting a time limit on the authority it confers, but Congress's failure to designate a termination date does not necessarily prove that Phase Three might last forever. There is no general principle in American public law by which a statute becomes obsolescent, but a law can be so tied to a particular situation that the passage of time and change of circumstances might make it impossible for the executive branch to rely in good faith on the law for legal authority. The AUMF was not a blank check empowering the executive to war on any and all potential enemies in perpetuity, and at some point it may no longer be plausible to assert that the terrorists of today are in continuity with the perpetrators of 9/11 in any significant way. In a major 2013 speech, President Obama observed that "[n]either I, nor any President, can promise the total defeat of terror," but he went on to say that "[c]ore al Qaeda is a shell of its former self. Groups like AQAP must be dealt with, but in the years to come, not every collection of thugs that labels themselves al Qaeda will pose a credible threat to the United States," or (he might have added) have any genuine claim to being part of the original organization that carried out the 9/11 attacks. An inquiry into the constitutionality of killing al-Awlaki would not be complete without considering what temporal limits may exist to the government's claim of authority.

Phase Four. By the fall of 2014, three years after the killing of al-Awlaki, the Phase Three war against al-Qaeda was quickly becoming less prominent in public discussion than a new conflict, with a new organization, the Islamic State (IS), also known as ISIS (the Islamic State of Iraq and al-Sham) or ISIL (the Islamic State of Iraq and the Levant). The Islamic State emerged in the first decade of the twenty-first century in the wake of the invasions of

Afghanistan and Iraq, the Iraqi insurgency against American occupation, and the ongoing civil war in Syria. Key figures in the origins of IS had been part of al-Qaeda and other terrorist organizations, but the Islamic State is a significantly different type of entity.

> Bin Laden viewed his terrorism as a prologue to a caliphate he did not expect to see in his lifetime. His organization was flexible, operating as a geographically diffuse network of autonomous cells. The Islamic State, by contrast, requires territory to remain legitimate, and a top-down structure to rule it. (Its bureaucracy is divided into civil and military arms, and its territory into provinces.)[20]

Recognized as a legal government by no member of the United Nations, IS views itself as the restored Islamic caliphate, with worldwide authority, uses conventional military means to expand its territorial base, and functions as a government in areas that it controls.[21] As of spring 2015, according to some estimates, IS "rules an area larger than the United Kingdom."[22]

20. Graeme Wood, What ISIS Really Wants (Mar. 2015), http://www.theatlantic.com/features/archive/2015/02/what-isis-really-wants/384980/.

21. *See* [Anonymous], The Mystery of ISIS, http://www.nybooks.com/articles/archives/2015/aug/13/mystery-isis/:

> In 2011, after the US withdrawal [from Iraq], the new leader, Abu Bakr al-Baghdadi, expanded into Syria and reestablished a presence in northwest Iraq. In June 2014 the movement took Mosul—Iraq's second-largest city—and in May 2015 the Iraqi city of Ramadi and the Syrian city of Palmyra, and its affiliates took the airport in Sirte, Libya. Today, thirty countries, including Nigeria, Libya, and the Philippines, have groups that claim to be part of the movement. . . . [I]n some occupied cities such as Raqqa in Syria, the movement has created complicated civil service structures, taking control even of municipal waste departments. [T]he revenue it derives from local income and property taxes, and by leasing out former Iraqi and Syrian state offices to businesses . . . has given ISIS a broad and reliable income base.

22. Wood, *supra* note 20.

Perceiving the Islamic State as a serious threat to U.S. national interests, one of "the radical groups [that] exploit grievances for their own gain" and constitute "the greatest threats" to "the security of the American people,"[23] in August 2014, the executive branch began airstrikes against IS, with targeted attacks on senior IS leaders an avowed tactic.[24] The extent to which the United States is relying on drone strikes is unclear but apparently expanding.[25] By all accounts the level of U.S. military action, albeit in the asymmetrical form of air power against ground forces, has been substantial: hundreds of air attacks launched, thousands of IS fighters killed. The United States sharply contests the claims of many observers that the strikes have caused substantial civilian casualties as well.[26] The chief of staff of the U.S. Army has publicly questioned statements by other officials that military efforts against IS might take three to five years: In General Ray Odierno's view, "ISIS is a ten to twenty year problem."[27]

23. Statement by the President on ISIL (Sept. 10, 2014), https://www.whitehouse.gov/the-press-office/2014/09/10/statement-president-isil-1.

24. *See* Air Campaign Against ISIS to Target Leaders and Seized U.S. Weapons (Sept. 12, 2014), http://www.nbcnews.com/storyline/isis-terror/air-campaign-against-isis-target-leaders-seized-u-s-weapons-n200861.

25. One Year On, Drone Attacks Against ISIS Increasing (Aug. 8, 2015), http://www.cbc.ca/news/world/one-year-on-drone-attacks-against-isis-increasing-1.3182002.

26.

> Between August 8th 2014 and June 30th 2015, the United States and its eleven Coalition allies had already carried out 4,924 airstrikes against Islamic State/Daesh across both Iraq and Syria. Almost 17,000 munitions were released . . . Official estimates of the number of Daesh fighters killed in this period ranged from 10,000 to 13,000. . . . Airwars believes there is sufficient publicly-available evidence to indicate Coalition responsibility for . . . 459–591 alleged civilian fatalities [in contrast to the] Coalition's admission of only two "*likely*" non-combatant deaths to date. . . . US and allied claims of almost zero civilian deaths are, in our view, untenable.

Airwars, *Cause for Concern: Civilians Killed in Coalition Strikes* 6, 5, 8 (Aug. 2015).

27. Odierno: ISIS Fight Will Last "10 to 20 years" (July 17, 2015), http://www.defensenews.com/story/defense/2015/07/17/odierno-isis-fight-last-10-20-years/30295949/.

Despite the evident differences between a terrorist network such as al-Qaeda and a territorially-based entity, the Obama administration's rhetorical position has been to assimilate IS to earlier enemies. In explaining his decision to attack IS, for example, President Obama asserted that the Islamic State "is certainly not a state. . . . ISIL is a terrorist organization, pure and simple"[28] Consistent with this view of the enemy, administration officials have invoked the September 2001 AUMF (and also the 2002 authorization for the invasion of Iraq) as an adequate basis for the air war since Congress seems unlikely to pass a new AUMF to cover U.S. actions against IS.[29] The existence of continuing operations aimed at al-Qaeda targets enhances the surface plausibility of treating IS-directed military operations as covered by the same legislative authorization but this is more a matter of appearance than political or military reality.[30] Reliance on the earlier statutes is problematic at best: The Islamic State did not even exist in September 2001 and seems more a competitor with al-Qaeda

28. Statement by the President on ISIL, *supra* note 23.

29. *See* Push to Write New War Powers for Obama Stalls in Congress (May 18, 2015), http://www.chicagotribune.com/news/nationworld/ct-obama-war-powers-20150518-story.html.

30. Over time, the public legitimacy of relying on Phase Three authorization has declined, even for operations against targets the executive branch genuinely believes to belong to al-Qaeda. The October 15, 2015, disclosure by the online magazine *The Intercept* of leaked, classified materials about drone strikes illustrates this point. As one commentator accurately observed, "[t]here actually isn't all that much new information" in the leaked materials, and for the purposes of this book nothing of constitutional significance was disclosed. See Quinta Jurecic, What the Intercept Found in "The Drone Papers"—And What I Found in them, https://www.lawfareblog.com/what-intercept-found-drone-papers%E2%80%94and-what-i-found-them (Oct. 16, 2015). It does seem fair to conclude that the materials sharpen concerns over how successful the U.S. government actually is in minimizing civilian casualties, but nothing in them appears to me to justify the assertion, by members of Congress among others, that the leaked information supports the conclusion that "[t]he U.S. drone program operates on highly questionable legal ground," at least from the standpoint of domestic law. Jenna McLaughlin, Congressman Calls for More Oversight After Drone Papers Revelations, https://theintercept.com/2015/

than an ally or affiliate,[31] and it has no connection with Saddam Hussein's long-defunct Iraqi government. But resting the lawfulness of the air war against IS directly on Article II raises serious questions about what standard (if any) the Obama administration is using for gauging the scope of unilateral presidential authority. As suggested above, would the expiration of AUMF authority simply return the constitutional situation to what I am calling Phase One? If that is the best analysis, the practical consequences for the executive would be both limiting (by its own legal tradition, the executive would be obligated to avoid military action rising to the level of "war") and liberating (the AUMF's definitions of *whom* the president is authorized to fight would no longer apply). Against that view, the factual continuities between the congressionally-authorized war against al-Qaeda and the new presidentially-initiated conflict with IS might be thought to invite a more aggressive use of military force against IS on the ground that doing so serves Congress's broader policy even if it not within the letter of Congress's actual legislation. These uncertainties suggest that the use of military force, including targeted killing by drone, against the Islamic State is best seen as a new Phase Four in our history, one that will test the plausibility of the arguments I present about the constitutionality of the decision to kill Anwar al-Awlaki. It is time to begin that presentation.

10/23/congressman-calls-for-more-oversight-after-drone-papers-revelations/ (Oct. 23, 2015) (quoting Rep. Keith Ellison (D-Minn.). The materials are exclusively concerned with drone strikes directed at what the executive appears to have believed al-Qaeda targets.

31. The two organizations appear to combine fundamental hostility toward the other's legitimacy with a potential willingness to engage in temporary and tactical cooperation. *See, e.g.*, http://www.reuters.com/article/2015/09/09/us-mideast-zawahri-idUSKCN0R91LY20150909#Z6k1tVUjGZKrjYHW.97 (Sept. 9, 2015) (quoting al-Qaeda leader Ayman al-Zawahri: "if I were in Iraq or Syria I would cooperate with [IS] in killing the crusaders and secularists and Shi'ites even though I don't recognize the legitimacy of their state").

Chapter 2

How to Think Constitutionally

If we are to determine whether it was constitutional for the government of the United States to kill Anwar al-Awlaki, we have to understand what it means to think constitutionally about the problem. Since the Constitution of the United States is a written document—and not, say, a set of conventions and usages as the British constitution is often said to be—it would be natural to assume that constitutional law is a straightforward matter of documentary interpretation, requiring nothing but that one "lay the article of the Constitution which is invoked beside the statute [or other governmental action] which is challenged and . . . decide whether the latter squares with the former."[1] As the introduction stated, this book presupposes a rather different understanding of

1. United States v. Butler, 297 U.S. 1, 62 (1936). The opinion in *Butler* that gave this deceptively simple description of the Supreme Court's duty in constitutional cases was itself a clear example of the description's inadequacy: In reaching the conclusion that the act of Congress before it was invalid, the Court found it necessary not only to construe the meaning of the words "the common Defence and general Welfare" in Article I, section 8 but also to decide whether a federal spending program can be unconstitutional because its effect is to coerce its "beneficiaries" to accept the federal funds. The Court's answer (yes) was perfectly sensible and (as a general principle) remains the law, but the Justices came to it by considerations of constitutional purpose and structure that went far beyond a semantic or historical interpretation of the Constitution's words.

what it means to reason through a legal problem that arises under the American Constitution. Constitutional thought *is* a matter of bringing the written Constitution to bear on the issues, but the Constitution "announc[es] fundamental principles in value-laden terms that leave ample scope for the exercise of normative judgment by those charged with interpreting and applying it."[2] Justifying this description of constitutional thought is a task for another day; this chapter's purpose, instead, is to lay out how we are think about constitutional issues if we assume that the Constitution addresses not only details about the mechanics of government (each state is represented by two senators, the president serves for a four-year term, and so on) but also "fundamental principles" about the norms and values of the American political community.

A. THINKING CONSTITUTIONALLY: REASONING ABOUT TEXT AND PURPOSES

Justice John Marshall Harlan II once wrote that "the basis of judgment" in determining the validity of governmental action under the Constitution of the United States

> must be a rational one, approaching the text . . . not in a literalistic way, as if we had a tax statute before us, but as the basic charter of our society, setting out in spare but meaningful terms the principles of government.[3]

2. Thornburgh v. Am. Coll. of Obstetricians & Gynecologists, 476 U.S. 747, 789 (1986) (White, J., dissenting), *overruled by* Planned Parenthood v. Casey, 505 U.S. 833 (1992).

3. Poe v. Ullman, 367 U.S. 497, 540 (1961) (Harlan, J., dissenting). Justice Harlan's dissent was later vindicated by the Court in the line of decisions beginning with *Griswold v. Connecticut*, 381 U.S. 479 (1965).

Harlan, a brilliant and deeply traditional lawyer, did not pretend to originality in this understanding of what it means to think constitutionally: He followed the words just quoted with a citation to the Supreme Court's decision in *McCulloch v. Maryland*, the greatest opinion written by the single greatest figure in the history of constitutional law, John Marshall, Chief Justice of the United States from 1801 to 1835.[4]

As a matter of history, Marshall's preeminence in U.S. constitutional law is beyond dispute: as Justice Benjamin Cardozo, himself a giant figure in the history of American law, once said, Marshall

> gave to the Constitution of the United States the impress of his own mind; and the form of our constitutional law is what it is, because he moulded it while it was still plastic and malleable in the fire of his own intense convictions.[5]

The constitutional convictions that Cardozo had in mind were, in part, Marshall's positions on the substance of constitutional law—above all, the overriding importance of national union as against the centrifugal forces of state parochialism—and, in part, Marshall's understanding of the appropriate processes of constitutional decision. For Marshall, the question of how we are to think constitutionally is no dry issue of methodology but goes to the very heart of the constitutional enterprise.[6]

4. 17 U.S. (4 Wheat.) 316 (1819).

5. Benjamin Cardozo, The Nature of the Judicial Process 169–70 (1921).

6. In a vindication of *McCulloch* written under a nom de plume, Marshall "assert[ed] that the reasoning is less doubtful than the conclusion. I myself concur in the conclusion; but I do not fear contradiction from any fair minded and intelligent man when I say that the principles laid down by the court for the construction of the constitution may all be sound, and yet the act for incorporating the Bank be unconstitutional." John Marshall, *A Friend to the Union II (April 28, 1819)*, *in* John Marshall's Defense of McCulloch v. Maryland 103 (Gerald Gunther ed., 1969).

For Marshall as much as for Justice Harlan, constitutional thought addresses "the principles of [American] government" as they are set out in "the basic charter of our society." The idea that one can faithfully resolve disputes over fundamental governmental principles as the exercise of a narrowly professional reason that leads to ineluctably correct conclusions is a radical misunderstanding, both of the sort of thinking involved and of the type of certainty one must seek. "The judgment is so much influenced by the wishes, the affections, and the general theories of those by whom any political proposition is decided," Marshall once wrote, "that a contrariety of opinion on [a] great constitutional question ought to excite no surprise."[7] The only answer—but Marshall thought it a wholly adequate answer—one should expect to a "great constitutional question" is a conclusion that makes sense of the Constitution in light of "that great paramount law of reason, which pervades and regulates all human systems."[8]

Marshall discussed the relationship between his view of constitutional reasoning and his unfeigned acceptance of the authority of the constitutional text in an essay he wrote defending his opinion in the *McCulloch* case:

> [The Constitution] is intended to be a general system for all future times, to be adapted by those who administer it, to all future occasions that may come within its own view. From its nature, such an instrument can describe only the great objects it is intended to accomplish, and state in general terms, the specific powers which are deemed necessary for

7. 4 JOHN MARSHALL, LIFE OF GEORGE WASHINGTON 243 (Chelsea House 1983) (1805).

8. John Marshall, *A Friend of the Constitution III, Alexandria Gazette (July 2, 1819)*, *in* JOHN MARSHALL'S DEFENSE, at 168.

> those objects. . . . It is impossible to construe such an instrument rightly, without adverting to its nature.[9]

For Marshall, and the mainstream of American constitutional thought after him, constitutional reasoning cannot be limited to a semantic or historical exegesis of the document's language. Such a severely textualist understanding of constitutional law is unfaithful to the very text it purports to interpret. If we are to find responsible answers to difficult constitutional questions—and bearing in mind that not all questions are difficult—we must think of the Constitution's purposes, its "great objects," and whenever possible give its words the interpretation that best achieves the document's purposes. Marshall recognized the possibility that on occasion the constitutional text may require a conclusion that seems less than optimal in light of its purposes. "If, indeed, such be the mandate of the constitution, we have only to obey."[10] But the general principle is that the written Constitution actually commands the outcomes that make sense in light of "the great paramount law of reason."

Chief Justice Marshall's understanding of constitutional thinking did not go unchallenged at the time any more than his strongly nationalist views on federalism escaped criticism: On both matters, his arch-rival was Thomas Jefferson, who argued for precisely the sort of rigorously text-bound thinking on constitutional questions that Marshall rejected. The appeal of Jeffersonian textualism, then and since, lies in its commendable ambition to be faithful to the Constitution, and it has counted among its adherents two of the most prominent Justices in the modern history of the Supreme Court, Hugo Black and Antonin Scalia. But the Jeffersonian position has never been the

9. *Id.* at 170–71.
10. *McCulloch*, 17 U.S. (4 Wheat.) at 408.

constitutional mainstream,[11] and even would-be Jeffersonians have usually found that in addressing difficult constitutional issues, "[t]he question is not whether the text shall be respected, but rather how one goes about respecting a text of that high generality and consequent ambiguity which marks so many crucial constitutional texts."[12] Marshall and Jefferson represent conflicting views on how best to respect the authority of the written Constitution, and so far in American history, the verdict in practice has been that Marshall's is the more faithful approach.[13]

The seminal statement of Chief Justice Marshall's approach to thinking constitutionally was, as Justice Harlan suggested, Marshall's opinion for the Court in *McCulloch v. Maryland*, decided in 1819. *McCulloch* addressed a burning constitutional issue in the early Republic, the constitutionality of a national bank; the Supreme Court held unanimously that the Constitution

11. In making this assertion, I do not intend to overlook or disparage the very considerable impact that Justice Scalia's repristination of Jeffersonianism has had on constitutional argument in recent decades. Advocates and judges alike are far more likely to make textualist and original meaning arguments today than was the case forty years ago. The significance of this shift in argumentative strategy is uncertain: Such arguments have been a staple of constitutional debate from the beginning, and in most areas of constitutional law the predominant mode of reasoning remains within the broader non-Jeffersonian framework that I am labeling mainstream. Most of the time Scalia himself works within that framework. *See, e.g.*, Maryland v. Shatzer, 559 U.S. 98 (2010) (Scalia, J., for the Court) (announcing a constitutional rule that a fourteen-day break in custody after a criminal suspect invokes his Fifth Amendment right to counsel will terminate the presumption that his subsequent waiver of the right to counsel in response to police interrogation was coerced).

12. Charles L. Black, Jr., Structure and Relationship in Constitutional Law 30 (1969).

13. United States v. Comstock, 560 U.S. 126 (2010), provides a striking example of the current Supreme Court's adherence to the Marshallian perspective. Four Justices filed opinions sharply disagreeing with one another over the appropriate analysis of an issue arising under the Necessary and Proper Clause. All four opinions purported to rely on *McCulloch* as exemplifying appropriate constitutional method; Jefferson, in contrast, appeared in only one, and even there in passing. *See id.* at 150 (Thomas, J., dissenting) (quoting a letter of Jefferson warning against "congressional powers becom[ing] completely unbounded by linking one power to another").

authorized Congress to charter the second national bank as a means of carrying out the federal government's fiscal operations.[14] To grasp fully the style of constitutional reasoning that Marshall's opinion employs—and urges the reader to accept—we should have in mind the contrasting Jeffersonian style that Marshall was clearly (if implicitly) rejecting. Marshall was thoroughly familiar with Jefferson's 1791 opinion on the validity of the first national bank bill:[15] His opinion in *McCulloch* was a sustained and self-conscious answer to Jefferson. If we understand the ways in which Marshall and Jefferson differed on how to think about constitutional issues, we will be better equipped to tackle an issue such as targeted killing where there is relatively little case law—Marshall and Jefferson, one should recall, were addressing the national bank question with essentially (in Jefferson's case, literally) no judicial precedent on which to rely.

In addressing the question whether Congress had the power to create a national bank, Jefferson presupposed that the answer could only be found through a clause-by-clause examination of each provision in Article I, section 8 that might be thought to afford Congress the power. In each instance, Jefferson relied on one or both of two factors in concluding that the particular clause did not authorize the bank bill: the semantic meaning of

14. *McCulloch* also ruled on (and against) the constitutionality of a state tax imposed on the bank.

15. *See* Thomas Jefferson, *Opinion on the Constitutionality of the Bill for Establishing a National Bank (February 15, 1791), in* JEFFERSON POWELL, LANGUAGES OF POWER: A SOURCEBOOK OF EARLY AMERICAN CONSTITUTIONAL HISTORY (1991). Before deciding to sign the first national bank bill into law, President Washington requested and received formal legal opinions from Secretary of State Jefferson, Secretary of the Treasury Alexander Hamilton, and Attorney General Edmund Randolph. Hamilton, who proposed the national bank, unsurprisingly advised Washington that it was constitutional, while Jefferson and Randolph concluded that the bill exceeded Congress's powers. Marshall published the Jefferson and Hamilton opinions in his biography of Washington. *See* 5 MARSHALL, *supra* note 7, appendix Note V.

the provision (the commerce clauses empower Congress "to prescribe regulations for buying and selling" not "to erect a bank") and the consequences of concluding that the provision did authorize a bank (if the bank were thought to be a valid, "necessary and proper" means to the collection of taxes, the Necessary and Proper Clause "would swallow up all the delegated powers"). This "clause-bound interpretivism" was the logical means of carrying out what Jefferson took to be the primary purpose of the Constitution, the imposition of limitations on a federal government that we cannot do without but that we cannot trust: He informed President Washington that "the foundation of the Constitution" is the principle of limited federal power announced in the tenth amendment and executed through Article I's grant of discrete and definable powers. "To take a single step beyond the boundaries thus specially drawn around the powers of Congress, is to take possession of a boundless field of power." Jefferson was thus doubly a constitutional pessimist: pessimistic about "those whom we are obliged to trust with power" and pessimistic about the capacity of common law reasoning to effectuate the Constitution by any means besides strict textual literalism. From this perspective, the semantic and consequential sides of his interpretive procedures were inextricably linked: That "single step beyond the boundaries" laid down by semantic limits of the text, Jefferson insisted, would immediately leave Congress with "boundless" power "no longer susceptible of any definition," while any argument that led to the consequence of expansive congressional power was invalidated by that very fact as an interpretation of the text.[16]

16. Jefferson, *supra* note 15, at 42–43. The phrase about the officials whom we must "trust with power" comes from Jefferson's 1798 draft resolutions on the unconstitutionality of the Alien and Sedition Acts. *See* Kentucky Resolutions of 1798, in POWELL, *id.* at 133. John Hart Ely coined the phrase "clause-bound interpretivism," which perfectly describes Jefferson's understanding of constitutional

When we turn to Marshall's opinion in *McCulloch*, we enter a different constitutional world: Marshall is optimistic (guardedly) about the federal government as the servant of the people and optimistic (robustly) about the capacity of common law reasoning to reach conclusions that make sense of the constitutional system. The equivalent passage in *McCulloch* to Jefferson's "foundation of the Constitution" discussion concludes with Marshall's famous admonition that "we must never forget that it is a *constitution* we are expounding," but the substance of Marshall's rhetoric lies in his rejection of Jefferson's insistence on parsing the constitutional text as if it were meant to provide in exhaustive detail exactly which specific sorts of legislation Congress is permitted to enact—nothing in the text itself, Marshall insisted, gives one any reason to assume that "everything granted [to Congress is] expressly and minutely described." The Constitution's "nature ... requires, that only its great outlines should be marked, its important objects designated, and the minor ingredients which compose those objects, be deduced from the nature of the objects themselves." This process of deducing constitutional details is not a matter of insisting on arbitrarily chosen dictionary definitions, but of discerning what makes sense of the overall constitutional enterprise: Far from supporting Jefferson's clause-bound interpretivism, the Tenth Amendment, like the original constitutional text, "leav[es] the question, whether the particular power which may become the subject of contest, has been delegated to the one government, or prohibited to the other, to depend on a fair construction of the whole instrument."[17]

law. Professor Ely thought such an approach impossible to carry out in practice. *See* John Hart Ely, Democracy and Distrust 11–14 (1980).

17. 17 U.S. (4 Wheat.) at 406–07.

In the paragraph immediately following the line about not forgetting that constitutional law is about a *Constitution* as Marshall understands that concept, he expands on the nature of *this* Constitution.

> The sword and the purse, all the external relations, and no inconsiderable portion of the industry of the nation, are entrusted to its government. It can never be pretended, that these vast powers draw after them others of inferior importance, merely because they are inferior. Such an idea can never be advanced. But it may with great reason be contended, that a government, entrusted with such ample powers, on the due execution of which the happiness and prosperity of the nation so vitally depends, must also be intrusted with ample means for their execution. The power being given, it is the interest of the nation to facilitate its execution. It can never be their interest, and cannot be presumed to have been their intention, to clog and embarrass its execution, by withholding the most appropriate means. Throughout this vast republic, from the St. Croix to the Gulf of Mexico, from the Atlantic to the Pacific, revenue is to be collected and expended, armies are to be marched and supported.[18]

The United States has a federal system rather than a unitary government (in Marshall's day the usual term was "consolidated"), and so governmental powers and duties are delegated to the national government. But those powers and duties that are national include, let us note, "the sword" and "all the external relations" and, as Marshall says later on, include generally "those

18. *Id.* at 408.

great powers on which the welfare of a nation essentially depends." Powers of this nature are not to be understood in a parsimonious or grudging fashion: The very purpose of the Constitution is to empower the federal government to address "[t]he exigencies of the nation" and to impose on that government the duty to do so by the most effective and appropriate means.[19]

This purposive approach to constitutional reasoning implies no disrespect for the authoritative text, Marshall claims—in constitutional law, "general reasoning" cannot properly lead to a conclusion that is contrary to the "fair construction of the whole instrument." "If, indeed, such be the mandate of the constitution, we have only to obey." But "unless the words imperiously require it," it is erroneous constitutional reasoning to deny the validity of federal government action that carries out the duties for which the Constitution "grant[ed] these powers for the public good."[20] The basic point of the text itself is to create a constitutional system that will maintain its essential characteristics over time while serving the changing needs of society as those needs are perceived by those temporarily charged with responsibility for the Republic.[21] For Jefferson, constitutional reasoning is the search for the meaning of the words which the people adopted to protect themselves from government, and constitutional

19. *Id.* at 415, 408. *See id.* at 409–10: "The government which has a right to do an act, and has imposed on it, the duty of performing that act, must, according to the dictates of reason, be allowed to select the means; and those who contend that it may not select any appropriate means, that one particular mode of effecting the object is excepted, take upon themselves the burden of establishing that exception."

20. *Id.* at 411, 407, 408.

21. The Constitution is "intended to endure for ages to come, and consequently, to be adapted to the various crises of human affairs.... we think the sound construction of the constitution must allow to the national legislature that discretion, with respect to the means by which the powers it confers are to be carried into execution, which will enable that body to perform the high duties assigned to it, in the manner most beneficial to the people." *Id.* at 415, 421. Professor Jamal Greene points out that in this passage and elsewhere in *McCulloch* Marshall refuses to give

reasoning can proceed along algorithmic lines. For Marshall, constitutional reasoning is the search for the decision that will best serve the purposes of the government the people created to serve their needs. The text's authority is incontestable, but the text provides us with no algorithm that can obviate the need to exercise judgment in difficult constitutional issues.

In *McCulloch*, Chief Justice Marshall's focus was on the scope of congressional authority in the context of a claim that Congress was exceeding its constitutional reach and invading a domain of state authority. The case raised no questions about the relationship between the powers of Congress and those of the president. The seminal opinion on the constitutional relationship between the two political branches of the federal government, an opinion very much in the Marshall tradition, was Justice Robert H. Jackson's concurrence in the much-later Steel Seizure Case, *Youngstown Sheet & Tube Co. v. Sawyer.*[22] *Youngstown* involved a

a narrow Jeffersonian reading to the term "necessary" in the Necessary and Proper Clause because he maintains that doing so would misunderstand the purposes of the constitutional text.

> [T]he argument for a particular reading of the text follows from the fact that "those who gave these powers" had particular aims in mind. It does not follow directly, in the originalist sense that might simply assimilate the meaning of the word "necessary" to the intentions of its authors. Rather, it follows in the sense that a contrary reading would disappoint the Founders and diminish their work: It would "give [the Constitution] the properties of a legal code." Respect for the "character of the instrument" therefore pushes toward a particular reading of its language. A narrow reading is beneath the U.S. Constitution.

Jamal Greene, *Pathetic Argument in Constitutional Law*, 113 COLUM. L. REV. 1389, 1426–27 (2013) (quoting McCulloch v. Maryland, 17 U.S. (4 Wheat.) at 415).

22. 343 U.S. 579 (1952). On the authoritative force of the Jackson opinion in analyzing the constitutionality of executive action, *see* Zivotofsky ex rel. Zivotofsky v. Kerry, 135 S. Ct. 2076, 2083 (2015) ("In considering claims of Presidential power this Court refers to Justice Jackson's familiar tripartite framework from *Youngstown*"); Medellin v. Texas, 552 U.S. 491, 524 (2008) ("Justice Jackson's familiar tripartite scheme provides the accepted framework for evaluating executive action").

Korean War–era challenge by U.S. steel companies to President Harry S. Truman's seizure of the industry for the stated purpose of avoiding a strike by the industry labor unions. A divided Supreme Court held that Truman had acted unlawfully in ordering the seizure despite the executive's vigorous argument that the action was necessary in order to prevent a steel shortage that could be disastrous for the U.S. war effort. Each of the six Justices in the majority wrote a separate opinion, and although Jackson wrote only for himself, and indeed joined Justice Black's official opinion of the Court, Jackson's concurrence has now achieved canonical status as a statement of separation of powers law. His opinion also exemplifies the Marshallian understanding of how to think constitutionally applied to a context rather closer to our question than the old dispute over the national bank.

In his first paragraph, Jackson described himself as a former "legal adviser to a President in time of transition and public anxiety" whose subsequent service on the Court had provided him a chance for "detached reflection" without fundamentally changing his sense of the issues: Much of his opinion can be read as advice for future executive branch lawyers dealing with novel issues of presidential authority in the absence of extensive case law and in light of the ambiguities of text and history that are especially pronounced with regard to the scope of the president's powers.[23] Jackson echoes Marshall in *McCulloch* in two

23. 343 U.S. at 634 (Jackson, J., concurring).

A judge, like an executive adviser, may be surprised at the poverty of really useful and unambiguous authority applicable to concrete problems of executive power as they actually present themselves. Just what our forefathers did envision, or would have envisioned had they foreseen modern conditions, must be divined from materials almost as enigmatic as the dreams Joseph was called upon to interpret for Pharaoh. A century and a half of partisan debate and scholarly speculation yields no net result but only supplies more or less apt quotations from respected sources on each

important respects. Like Marshall, Jackson rejects Jefferson-style clause-bound interpretivism: "The actual art of governing under our Constitution does not and cannot conform to judicial definitions of the power of any of its branches based on isolated clauses or even single Articles torn from context." And again like Marshall, Jackson argues that in the absence of much guidance from "the conventional materials of judicial decision," a constitutional lawyer must seek conclusions that make the most sense of the overall constitutional enterprise.[24]

Youngstown, however, was unlike *McCulloch* in that the Court's direct concern was not with the national government's overall adequacy to its great purposes, but with the constitutional commitment to achieving those purposes through a system of distinct governmental institutions that protects liberty by dispersing power without disabling it.

> [A]s we approach the question of presidential power, we half overcome mental hazards by recognizing them. The opinions of judges, no less than executives and publicists, often suffer the infirmity of confusing the issue of a power's validity with the cause it is invoked to promote, of confounding the permanent executive office with its temporary occupant. The tendency is strong to emphasize transient results upon policies . . . and lose sight of enduring consequences upon the balanced power structure of our Republic. . . . While the Constitution diffuses power the better to secure liberty, it also contemplates that practice will integrate the dispersed powers into a

side of any question. They largely cancel each other. And court decisions are indecisive because of the judicial practice of dealing with the largest questions in the most narrow way.

Id. at 634–35.

24. *Id.*

> workable government. It enjoins upon its branches separateness but interdependence, autonomy but reciprocity.[25]

Sound constitutional thinking about the president's exercise of power is not solely a matter of asking whether the executive action is intended to further one of the Constitution's purposes. We must also take into account the Constitution's creation of a government that achieves its purposes through particular institutional arrangements that structure and limit what each branch of government may legitimately do.

Another important theme in Justice Jackson's Steel Seizure opinion is his insistence that if we are to think *constitutionally* about the president's authority, especially in circumstances of fear and crisis, we must strive to think with the concepts and categories found in the text and legal tradition and resist the temptation to import emotionally charged but legally indeterminate rhetoric into our analysis. The tripartite framework of analysis for which Jackson's opinion is most famous reflects his recognition that the language of Article I, section 8 and Article II, sections 2 and 3 identifies overlapping powers and responsibilities and does not justify arguments for presidential authority that amount to a claim of extralegal prerogative.[26] "Loose

25. *Id.*
26.

> 1. When the President acts pursuant to an express or implied authorization of Congress, his authority is at its maximum, for it includes all that he possesses in his own right plus all that Congress can delegate. . . .
>
> 2. When the President acts in absence of either a congressional grant or denial of authority, he can only rely upon his own independent powers. . . .
>
> 3. When the President takes measures incompatible with the expressed or implied will of Congress, his power is at its lowest ebb, for then he can rely only upon his own constitutional powers minus any constitutional powers of Congress over the matter.

Id. at 636–37.

and irresponsible use of adjectives colors all non-legal and much legal discussion of presidential powers. 'Inherent' powers, 'implied' powers, 'incidental' powers, 'plenary' powers, 'war' powers and 'emergency' powers are used, often interchangeably and without fixed or ascertainable meanings."[27] The president's authority as commander in chief should be "measured by the command functions usual to the topmost officer of the army and navy . . . His command power is not such an absolute as might be implied from that office in a militaristic system but is subject to limitations consistent with a constitutional Republic whose law and policy-making branch is a representative Congress."[28] Accordingly, much of Jackson's opinion is devoted to a careful examination of each basis in the constitutional text that the executive invoked as authority for President Truman's seizure of the steel mills.

Just as Marshall did in *McCulloch* with the Necessary and Proper Clause and the Tenth Amendment, Jackson examines a supposed tension between the relevant constitutional purpose (here, the maintenance of the balanced power structure) and provisions in the text (various clauses in Article II claimed to license extraordinary executive power) and concludes that when properly read the provisions are perfectly consistent with the purpose. Precisely because part of the point of the constitutional enterprise is to create and maintain a "balanced power

27. *Id.* at 646–47. I have hesitated to use the term "war powers" in light of Jackson's disapproval of the term, but in the end convenience won out. However, following the substance of Jackson's concern, in this book "war powers" refers specifically to Article I's enumeration of legislative powers relating to war, Article II's designation of the president as commander in chief, and the *McCulloch* principle found in the Necessary and Proper Clause that Congress may enact legislation that is "plainly adapted" to legitimate constitutional purposes. *See McCulloch*, 17 U.S. (4 Wheat.) at 421.

28. 343 U.S. at 645–46.

structure" that protects liberty even as it pursues the nation's other interests, Jackson thought that issues involving "the war power, which I regard as the Achilles Heel of our constitutional system," demand the greatest effort to avoid inexact language that might serve to undermine that structure.[29]

Justice Jackson's insistence on legal precision in defining the constitutional source for presidential action is equally applicable when analyzing the legitimacy of Congress's exercise of its powers.[30] As we shall see in Chapter Three, Article I's delegation of the power "to declare War" empowers Congress not simply to announce a state of total war, but to make decisions in general about the initiation and scope of military action. But the term "war" is not a metaphor, or an empty semantic container into which one can put whatever content may be convenient: Congress cannot declare "war" on drugs or crime, or perhaps even "international terrorism" in the abstract, and rely on the Declaration of War Clause for constitutional authority to pass legislation. To be within the scope of the power conferred by the clause, congressional action must relate to the use of military force against designated foreign enemies who can be fought by such means.[31]

29. The "Achilles Heel" quotation is from a concurring opinion that Justice Jackson drafted but ultimately did not file in *Hirabayashi v. United States*, 320 U.S. 81 (1943). The opinion is published, and the quotation to be found, in Dennis J. Hutchinson, *The Achilles Heel" of the Constitution: Justice Jackson and the Japanese Exclusion Cases*, 2002 SUP. CT. REV. 455, 468.

30. As Justice Jackson himself recognized. *See* Woods v. Cloyd W. Miller Co., 333 U.S. 138, 146 (1948) (Jackson, J., concurring) (expressing concern about the executive's invocation of a "vague, undefined and undefinable 'war power'" as the constitutional basis for an act of Congress).

31. The Supreme Court held during the Civil War that Congress cannot declare war against a state or group of states; the Civil War's constitutional legitimacy stemmed from the secessionists' use of force against the United States, which thereby forced war on the federal government. *See* Prize Cases, 67 U.S. 635 (1863).

B. THINKING CONSTITUTIONALLY: THE DISTINCTION BETWEEN AUTHORIZATIONS AND PROHIBITIONS

The Constitution plays two fundamental and distinct roles with respect to the U.S. government. It creates that government and distributes the powers it delegates among the three branches, and it puts in place specific prohibitions on the use of the powers it delegates. Congress, the executive, and the federal courts make up a limited government, therefore, in two distinct senses: The federal branches can lawfully exercise only certain governmental powers (and not others), and they exercise those powers subject to further limitations on how they employ their powers. This distinction between questions of constitutional authorization and questions of constitutional prohibition is fundamental and affects the proper analysis of many constitutional problems, including targeted killing. In *National Federation of Independent Business v. Sebelius*, the 2012 Affordable Care Act case, Chief Justice John Roberts succinctly explained the difference.

> Today, the restrictions on government power foremost in many Americans' minds are likely to be affirmative prohibitions, such as contained in the Bill of Rights. These affirmative prohibitions come into play, however, only where the Government possesses authority to act in the first place. If no enumerated power authorizes Congress to pass a certain law, that law may not be enacted, even if it would not violate any of the express prohibitions in the Bill of Rights or elsewhere in the Constitution.[32]

32. 132 S. Ct. 2566, 2577 (2012).

It is always true of any branch of the federal government that it "must [be able to] show that a constitutional grant of power authorizes each of its actions." If and only if a showing of authorization can be made, is it logically necessary to ask the second question whether an "affirmative prohibition, such as contained in the Bill of Rights," forbids the action in question.

The relationship between authorization and prohibition is not commutative: "[A]ffirmative prohibitions come into play . . . only where the Government possesses authority to act in the first place," and constitutional prohibitions override constitutional authorizations.[33] A good example is Congress's authority over immigration: Congress, the Supreme Court has consistently held, possesses " 'plenary power' to create immigration law," but its power to do so is subject "to important constitutional limitations [and it must choose] a constitutionally permissible means of implementing" that power that does not violate any applicable prohibition.[34]

The distinction between issues of authorization and prohibition, and the superiority of prohibitions over authorizations, are inferences from the constitutional text, which includes provisions clearly authorizing one or more of the branches to act and other provisions that prohibit certain kinds of action regardless of whether the actions are otherwise authorized: Think of Article I, section 8 and section 5 of the Fourteenth Amendment (authorizations of congressional action) as opposed to Article I, section 9 and the First Amendment (prohibitions on congressional

33. *Id.* at 2577–78. "The same [requirement that the federal government show that it 'possesses authority to act in the first place'] does not apply to the States, because the Constitution is not the source of their power. The Constitution may restrict state governments—as it does, for example, by forbidding them to deny any person the equal protection of the laws. But where such prohibitions do not apply, state governments do not need constitutional authorization to act." *Id.* at 2578.

34. Zadvydas v. Davis, 533 U.S. 678, 695 (2001) (citations omitted).

action). During the debate over ratification, the Constitution's proponents responded to anti-federalist complaints that the text lacked a bill of rights protecting such liberties as freedom of the press by arguing that these liberties were safe from federal interference because interference was not authorized expressly: "[T]he Constitution did not initially include a Bill of Rights at least partly because the Framers felt the enumeration of powers sufficed to restrain the Government. As Alexander Hamilton put it, 'the Constitution is itself, in every rational sense, and to every useful purpose, A BILL OF RIGHTS.'"[35] Freedom of the press is beyond congressional interference, on this reading, because Congress is nowhere authorized to interfere with it.

The Constitution's critics were not persuaded, however, and the constitutional text itself undercuts Hamilton's argument by including "affirmative prohibitions" in Article I, section 9 that would have been entirely redundant if the actions they forbade were clearly beyond anything authorized by the Constitution's grants of power. In the end, what we call the Bill of Rights was added to the Constitution, and after occasional hesitations in the founding era, the distinction between the limiting force of the Constitution's authorizations of some powers but not others, and the distinct limiting force of its prohibitions became settled principle. Thus a federal action may be both authorized (in the sense that it is within the scope of a delegated power) and yet prohibited. A simple example may clarify the point: Congress may provide for housing for the "Armies" it is authorized to "raise and support," but it may not do so in peacetime by requiring private homeowners to provide quarters for soldiers whether they wish to or not. The Third Amendment prohibition—"No Soldier shall, in time of peace be quartered in any house, without the

35. *Sebelius*, 132 S. Ct. at 2578 (quoting THE FEDERALIST No. 84).

consent of the Owner"—"trumps" the Article I, section 8, clause 12 authorization.

The analytical priority of constitutional prohibitions over constitutional authorizations does not mean that government may never take actions that seem at first glance to infringe a prohibition. Some prohibitions include express self-limitations: The Third Amendment, for example, expressly permits Congress to authorize the quartering of troops in private houses "in time of war ... in a manner to be prescribed by law." As the Supreme Court interprets them, other constitutional prohibitions are more limited than they appear to be as a semantic matter. The Court has always declined to construe the First Amendment, for example, as an absolute ban on all laws "abridging the freedom of speech" even though the amendment's text begins "Congress shall make no law." But that is, properly speaking, a matter of construing the scope of the prohibition on abridging speech, in the light of the Constitution's purposes, not a conclusion that at times Congress is entitled to violate freedom of speech.

There is no difference in terms of lawfulness between a federal action that is unconstitutional because the actor has no constitutional authorization to do it and an action that is unconstitutional because a constitutional prohibition bars it: In either case, the action is unlawful. But the authorization/prohibition distinction is vitally important in analyzing *whether* a federal action is constitutional. In the mainstream Marshallian tradition, as the reader knows, sound constitutional thought looks to constitutional purposes in addressing a difficult issue. In thinking about whether the Constitution authorizes some contemplated or contested action—the targeted killing of a U.S. citizen who is a major figure in al-Qaeda, let us say—we should give great weight to the reasonable implications of the purposes of the relevant constitutional grants of power to Congress and the president. If the conclusion to that inquiry is "yes, the killing is authorized," sound constitutional thought will then turn to the purposes of the relevant constitutional prohibition(s) in determining whether they bar Congress from approving or the executive from carrying out the killing.

The role of past political practice in our constitutional reasoning may change as well. The political branches are "expert" in the pursuit of the purposes for which they are delegated authority in two senses. First, as *McCulloch* expressly stated of Congress, the only reasonable view is that the Constitution grants Congress and the president wide discretion in using their powers to achieve the purposes of those powers. Second, the issues raised by constitutional authorizations are, much of the time, inextricably linked to or even dependent on judgments about the "real world" context and effects of the contemplated or challenged federal action; in making those judgments, expertise in the law is of little relevance. Whether al-Awlaki was the equivalent of Admiral Yamamoto for the purposes of targeting him with lethal military force is a *military and political* judgment, not a legal one, or at least it is heavily dependent on military and political analysis. The historical practice of military and political decision makers on such matters provides an interpretation in practice of the powers they are authorized to wield that is not driven by the specific issues of the day. In contrast, the very purpose of a constitutional prohibition is to override what may be an eminently sound view on the practical value of using a delegated power in a certain fashion by imposing a normative judgment about the compatibility of the federal action with fundamental principles. Whether al-Awlaki's U.S. citizenship ought to insulate him from a use of military force that could otherwise lawfully target him is a *legal* judgment about the meaning of U.S. citizenship, and the relevant forms of analysis are those of constitutional lawyers. Making such judgments is the central task of constitutional thinking, and it is the history of reasoned legal discussion, not of military or political decision, that is relevant.

The constitutional problem posed by the targeted killing of Anwar al-Awlaki requires both forms of constitutional analysis. Chief Justice Roberts wrote in the *Sebelius* case, "the restrictions on government power foremost in many Americans' minds are likely to be affirmative prohibitions, such as contained in the Bill of Rights." The public debate over the al-Awlaki killing bears out that judgment, and even the Obama administration's lawyers seem to have been far more concerned to rebut arguments that the killing violated the prohibition imposed by the Due Process Clause of the Fifth Amendment than to establish a clear constitutional authorization for the killing. Sound constitutional thinking will prevent us from slighting the question of authorization, or confusing the two issues.

C. THINKING CONSTITUTIONALLY ABOUT WAR

The text of the Constitution plainly divides responsibility for decisions about war and the use of military force between two political branches: Article I, section 8 delegates to Congress a variety of powers on the subject, and Article II, section 2 makes the president commander in chief of the nation's armed forces. The constitutional text does *not* explain the scope of the two branches' respective powers, or how to resolve possible conflicts in their exercise. There are, however, two constitutional principles of general application that will take us some way toward understanding how to relate the powers of Congress and the president over war in thinking about specific constitutional problems related to war.

The first principle is that as an ordinary matter we can analyze properly a question about the president's constitutional authority to take some action only if we take into account what Congress has done—and not done—that might bear on the action.[36] The classic statement of this principle comes from Justice Jackson's opinion in the Steel Seizure Case: "Presidential powers are not fixed but fluctuate, depending upon their disjunction or conjunction with those of Congress."[37] Taking account of the history of the war on terror that we briefly surveyed in Chapter One is essential to sound constitutional thought about the constitutionality of targeted killing (or anything else) as a means of prosecuting that war. In Phases One and Two, before Congress had legislated on the subject, the president could rely only on whatever authority Article II provides the executive to take unilateral action in those very different factual circumstances. Once Congress authorized the use of military force and thus Phase Three began, the president's constitutional authority

36. The qualifier in this sentence ("as an ordinary matter") is necessary because Article II vests the president with unconstrainable discretion in a small number of matters. Congress has no power, for example, to circumscribe the president's authority to issue a pardon or veto a bill. This does not mean that there are no constitutional limitations on the president in the exercise of those powers. Like all constitutional authorizations, the pardon and veto powers extend only to actions that *are* definitionally, as a constitutional matter, pardons and vetoes. Furthermore, in my view the president him- or herself is bound by the oath of office to respect constitutional prohibitions in exercising the pardon or veto power. I think it clearly would violate the First Amendment for the president to pardon one co-defendant and refuse to pardon the other because the president disapproved of the latter's religious beliefs; if the president were to do so, in my view that would open the president to impeachment and removal from office for a constitutional "high Crime." U.S. CONST. art. II, § 4 (defining grounds for impeachment). But neither Congress nor the federal courts could take action to prevent or nullify the first defendant's pardon or to compel the president to pardon the second defendant.

37. *Youngstown*, 343 U.S. at 635 (Jackson, J., concurring).

was shaped by the Authorization for the Use of Military Force: As we shall see in Chapter Three, Congress can provide the executive with options that Article II by itself does not afford, and by the same token Congress can place constraints on the president's choices that did not exist in the absence of legislation. A situation like the present Phase Four, with ongoing hostilities that no longer fit well within the existing congressional authorization, presents especially difficult constitutional issues that we shall examine in Chapter Five.

A second constitutional principle of general applicability that is especially relevant in thinking constitutionally about war is the rule that Congress acts on the world outside its four walls only by enacting legislation and therefore Congress lacks the authority to execute the powers of the federal government, including its own.[38] Under this rule, Congress could not have undertaken the targeting of al-Awlaki or ordered the armed forces to carry out the killing, even if it possesses the power to authorize or even oblige the president to do so.[39] Last-in-time responsibility for any decision to use military force is therefore, as a constitutional matter, always the president's. For that reason we shall orient our examination of federal war powers from the perspective of the executive branch.

38. On the principle, *see, e.g.*, Bowsher v. Synar, 478 U.S. 714, 726 (1986) ("The structure of the Constitution does not permit Congress to execute the laws").

39. We can leave to one side Article I, section 9, which specifically prohibits "bills of attainder," legislation "convicting" someone of a crime and prescribing his punishment. The bill of attainder issue was raised in the lawsuit brought after al-Awlaki's death, but the district court held that "[b]ecause Plaintiffs can point to no legislative action, the Bill of Attainder Clause does not apply." Al-Aulaqi v. Panetta, 35 F. Supp. 3d 56, 82 (D.D.C. 2014). My own view is that a statute purporting to require an act of war against a specific person would not implicate the clause because it would not be "*legislative action* intended to punish an individual." *Id.* at 82. The issue is theoretical since, for reasons that will become apparent, military decisions of that specificity are exclusively executive and thus the statute would be invalid on separation of powers grounds.

One can well imagine an objection at this point that our thinking is veering off into the stratosphere of purely academic issues that lack a purchase on reality. When the issue arises about whether to use military force, which always poses the possibility and (usually) the practical certainty that human beings will die, surely the perspective of "the Executive Branch of Government, whose particular responsibility is to maintain security" ought to be directed toward graver issues than satisfying the lawyer's professional taste for dotting i's and crossing t's.[40] The constitutional response to this objection is both yes and no. Undertaking the inevitable harms and potential risks of war is the most serious decision a political community can make, and the taking of human life has weighty ethical ramifications from any serious moral perspective. Lawyers, when speaking as lawyers, have little to contribute on these extralegal considerations; one of the central themes in the tradition of constitutional thought that began with *McCulloch v. Maryland* is that the Constitution of the United States generally places the responsibility for great political and ethical decisions in the hands of those who are politically and ethically responsible to the community. It was the opponents of the national bank, whose arguments Chief Justice Marshall rejected in *McCulloch*, who proposed a broader role for legal argument. The president should not look to his or her legal advisors for judgments about whether to order a particular military action.

At the same time, the Republic for which the president has "particular responsibility to maintain security" is a community of laws, constituted in the most literal sense by the Constitution that authorizes governmental action. Chief Justice John Roberts

40. Hamdi v. Rumsfeld, 542 U.S. 507, 545 (2004) (Souter, J., concurring in part, dissenting in part, and concurring in the judgment).

wrote for the Supreme Court a few years ago that even "compelling [governmental] interests . . . do not allow us to set aside first principles. The President's authority to act, as with the exercise of any governmental power, 'must stem either from an act of Congress or from the Constitution itself.' "[41] The executive branch has recognized this "first principle" since the administration of George Washington[42] and continues to do so, expressly, in the modern era. "The President's office and authority are created and bounded by the Constitution; he is required to act within its terms. Put somewhat differently, in serving as the executive created by the Constitution, the President is required to act in accordance with the laws—including the Constitution, which takes precedence over other forms of law."[43] The president, in short, has no legitimate power to act in the absence of legal authorization, and no legal source of authority except those created by

41. Medellin v. Texas, 552 U.S. 491, 524 (2008) (quoting *Youngstown*, 343 U.S. at 585). It is clear from Chief Justice Roberts's subsequent discussion that he did not intend to deny the theoretical possibility that the president could find domestic legal authority to act in a self-executing treaty. (A self-executing treaty is one that has domestic law force without requiring implementing legislation.) We may put that possibility to one side since there is no plausible argument that targeted killing by drone can be legally justified on the basis of any existing treaty. Given the view that tactical military decisions are exclusively presidential—a position advocated in Chapter Three and generally conceded—I assume that a treaty requiring the targeted killing of a specific individual would be an invalid infringement on the president's authority as commander in chief. To be sure, it is the *president* who makes treaties with the advice and consent of the Senate, but the Constitution does not permit a branch of the government to alienate its proper authority. *See, e.g.*, Free Enter. Fund v. Pub. Co. Accounting Oversight Bd., 561 U.S. 477, 497 (2010) ("the separation of powers does not depend on the views of individual Presidents nor on whether 'the encroached-upon branch approves the encroachment.' The President [cannot] choose to bind his successors by diminishing their powers.").

42. *See, e.g.*, H. Jefferson Powell, A Community Built on Words 11–21 (2002) (discussing Washington's solicitation of legal advice on the constitutional roles of the president and Senate in conducting diplomatic relations).

43. *Presidential Authority to Decline to Execute Unconstitutional Laws*, 18 Op. O.L.C. 199, 200 (1994).

and within the constitutional system.[44] Constitutional authorization for presidential action does not have to stem directly from Article II, of course, but can be mediated, the product of authority delegated to the executive by a constitutionally valid statute or treaty.[45] President Obama strongly reaffirmed the executive's commitment to this understanding of the constitutional system in a May 2013 speech: "Our laws constrain the power of the President, even during wartime, and I have taken an oath to defend the Constitution of the United States."[46] It is constitutional authority that legitimates the president's claim to act for the community, and a president who was indifferent to the existence of constitutional authority for a military decision would to that extent undermine the security of the United States beyond the capacity of any terrorist to harm the nation.

People sometimes hesitate over the statement that the president can act legitimately only on the basis of authority ultimately derived from the Constitution out of a worry that a situation might arise in which the president lacked the legal authority to

44. The mischievous suggestion to the contrary in *Curtiss-Wright Export Corp. v. United States*, 299 U.S. 304 (1936), is inconsistent with Chief Justice Roberts's first principle and generally (and rightly) discredited. *See* H. Jefferson Powell, *The Story of* Curtiss-Wright Export Corporation, *in* PRESIDENTIAL POWER STORIES 228–31 (Christopher H. Schroeder & Curtis A. Bradley eds., 2009) (discussing the decision's intellectual irrelevance).

45. Indeed, in the area of domestic governance, the president's authority *usually* stems from an act of Congress, since the president enjoys very few unilateral Article II powers that bear on domestic matters. The president has significant substantive Article II powers in the related areas of foreign affairs and national security. *See, e.g.*, Sale v. Haitian Ctrs. Council Inc., 509 U.S. 155, 188 (1993) ("foreign and military affairs" are those "for which the President has unique responsibility"). Even there, however, "Congress holds express authority to regulate public and private dealings with other nations in its war and foreign commerce powers," American Ins. Assoc. v. Garamendi, 539 U.S. 396, 414 (2003), and "[t]he Executive is not free from the ordinary controls and checks of Congress merely because foreign affairs are at issue." *Zivotofsky*, 135 S. Ct. at 2090.

46. Http://www.aclu.org/files/assets/obama-speech.pdf.

take some action that in his or her judgment was essential to national security. President Lincoln's decision to suspend habeas corpus in early 1861, and his implementation of that decision in the teeth of a judicial order by Chief Justice Taney, are the stock example of this danger.[47] But Lincoln's actions should occasion no doubts. First, Lincoln himself in a special message to Congress, and Attorney General Edward Bates in an elaborate legal opinion, defended the lawfulness of the executive's unilateral suspension order.[48] Second, and more important, as Lincoln himself acknowledged, his decision to suspend the writ was driven by his judgment that whatever the law might be, suspension was genuinely necessary to avert the danger that the secession movement would literally destroy the Republic if suspected secessionists could invoke the writ; executive suspension was necessary to save the very constitutional order in which the writ of habeas corpus has its force.[49] Given the precarious location of the national capital after Virginia purported to secede from the Union, with Washington, D.C., dependent on a single rail line across a Maryland apparently full of would-be secessionists, Lincoln's political and military judgment is certainly defensible. In a circumstance where a reasonable president can conclude in good faith that there is a threat to the survival of the nation or of

47. *See* Ex parte *Merryman*, 17 Fed. Cas. 144 (C.C.D.MD. 1861) (Taney, C.J., on circuit).

48. *See Suspension of the Privilege of the Writ of Habeas Corpus*, 10 Op. Att'y Gen. 74 (1861).

49. Abraham Lincoln, *Message to Congress in Special Session (July 4, 1861)*, *in* The Portable Abraham Lincoln 216 (Andrew Delbanco ed., 1992) ("are all the laws *but one* to go unexecuted and the Government itself go to pieces lest that one be violated? Even in such a case would not the official oath be broken if the Government should be overthrown, when it was believed that disregarding the single law [the guarantee of habeas corpus in Article I, section 9] would tend to preserve it?"). The Supreme Court's recent decision in *Boumediene v. Bush* may have resolved the constitutional disagreement between President Lincoln and Chief Justice Taney in Taney's favor. *See Boumediene*, 553 U.S. 723, 733 (2008).

its constitutional form of government, he or she might well conclude that there may be obligations deeper even than the president's high obligation to obey the law.[50]

Whatever one's views about this fascinating but perhaps unlikely scenario,[51] no one argues that the killing of al-Awlaki

50. Perhaps the most famous American statement of this principle is Jefferson's. *See* Thomas Jefferson, *Letter to John B. Colvin* (Sept. 20, 1810), *in* 3 THE PAPERS OF THOMAS JEFFERSON: RETIREMENT SERIES 99 (Susan Holbrook Perdue et al. eds., 2006) ("strict observance of the written laws is doubtless one of the high duties of a good citizen, but it is not the highest"). In their profound book on the roles of law and decency in the war on terror, Charles and Gregory Fried agree with Jefferson and Lincoln while pointing out that the president's constitutionally prescribed oath "does not mention defending national security; the president's duty is explicitly to the law, not some vague goal beyond the law." CHARLES FRIED & GREGORY FRIED, BECAUSE IT IS WRONG: TORTURE, PRIVACY AND PRESIDENTIAL POWER IN THE AGE OF TERROR 152 (2010). *See* U.S. CONST. art. II, § 1, cl. 8 ("I do solemnly swear (or affirm) that I will faithfully execute the Office of President of the United States, and will to the best of my Ability, preserve, protect and defend the Constitution of the United States."). Jefferson and the Frieds argued that the president's proper course if driven to act beyond the law is to submit his actions to public judgment, and Lincoln, of course, did exactly that. *See* Jefferson, *Letter to Colvin*, *id.* at 99 (an official who acts beyond his legal authority should "risk himself on the justice of the controlling powers of the Constitution," i.e., accept removal from office and condemnation if his judgment is rejected); FRIED & FRIED, *id.* at 154 ("a president who has acted against established law . . . should invoke the Constitution's ultimate judgment by inviting Congress to treat his lawbreaking as a 'high crime or misdemeanor'" subject to impeachment).

51. Professor Jack Goldsmith, a senior legal advisor during part of President George W. Bush's first term, has argued that the recent academic discussion of what is sometimes called the president's "prerogative power" to act beyond the ordinary law is essentially beside the point: In any genuinely imaginable situation, the president's lawyers would concoct a plausible enough legal argument to spare the president the politically unpalatable need to admit to lawbreaking. Jack Goldsmith, *The Irrelevance of Prerogative Power, and the Evils of Secret Legal Interpretation*, *in* EXTRA-LEGAL POWER AND LEGITIMACY: PERSPECTIVES ON PREROGATIVE 213–17 (Clement Fatovic & Benjamin A. Kleinerman eds., 2013). I have no quarrel with this assumption as a prediction about any likely administration's public pronouncements in such a situation, but it does not follow that the administration's claim of legal authority would be persuasive as a matter of sound constitutional thinking. *Cf.* Henry P. Monaghan, *The Protective Power of the Presidency*, 93 COLUM. L. REV. 1, 74 n.157 (1993) ("I accept the proposition that the government will act in an emergency when the dominant interests so require. And so, for me the question is whether in legal analysis the national government's emergency power is extra-constitutional.").

was an illegal action justified solely on the grounds that it was necessary to protect the nation: The Obama administration has consistently defended its targeted killing policies, like all of the measures it has taken in the war on terror, as lawful. And even if a future president were willing to contemplate breaking the law in order to save the nation, the first duty of the president's lawyers would be to determine just how far his or her lawful authority extends. The question of how to think constitutionally about the use of military force cannot be avoided on the ground that war may present issues deeper than legality.

Nor can the constitutional issue be evaded by characterizing actions such as the killing of al-Awlaki as the exercise of the executive's duty to enforce the criminal laws of the United States. There is a considerable literature on the question whether the U.S. government (and other governments) should address terrorists and terrorism in the legal and strategic framework created by the traditional concept of "war"; the alternative is to view terrorism as a particularly difficult issue of domestic and transnational law enforcement. Arguments can be made for either way of framing the issue: Twenty-first-century terrorists regularly use military-style weapons, lay claim to political legitimacy, and act in other respects as if they were soldiers in an armed if asymmetrical conflict rather than criminals; twenty-first-century terrorists are (almost invariably) guilty of crimes under U.S. and other law, wear no uniforms, do not respect international humanitarian law, and so on. The novel aspects of any "war on terror" have led some scholars to conclude that we need to devise an entirely distinct conceptual and legal framework because neither of the existing ones is entirely satisfactory; others argue that our institutional and legal arrangements must and in fact will adapt to the new strategic challenges posed by terrorism and that the only real choices are whether the adaptations combine strategic wisdom with fidelity to American constitutional values.[52]

For the purposes of this book, we need not resolve these difficult questions, because existing U.S. constitutional law and the actions of Congress and the president answer the framework issue: The existing U.S. struggle against terrorism of which the Obama administration's announced targeted killing policy is a part is an exercise of the federal government's authority to wage war, not of its (generally implied) powers to create and enforce criminal laws. The policy does not conform to the constitutional prohibitions governing criminal law enforcement in a variety of ways, and for that reason alone it has to be evaluated as a means of waging war rather than as a tool of law enforcement, unless we are at once to conclude that the policy is an unlawful violation of the relevant Bill of Rights prohibitions. But we do not need to infer whether the basic constitutional justification for the current policy is war or law enforcement since on that issue the administration has been quite clear. The various public statements by administration officials concur that "[t]he nature of this conflict . . . involves *actual* war—not war as a metaphor for policy seriousness, armed conflict in the strict legal sense. This is the US government's position even though the enemy is not a state."[53]

The Obama administration also defends the position that the war framework does not preclude it from using in appropriate settings and ways, a law enforcement paradigm and thereby relying on its ordinary criminal law resources. I argue in the next chapter

52. *See, e.g.*, PHILIP BOBBITT, TERROR AND CONSENT: THE WARS FOR THE TWENTY-FIRST CENTURY (2008); MARY KALDOR, NEW AND OLD WARS: ORGANIZED VIOLENCE IN A GLOBAL ERA (3d ed. 2012).

53. KENNETH ANDERSON & BENJAMIN WITTES, SPEAKING THE LAW 22 (2013-on).

that the administration is correct that this is constitutionally permissible.[54] But the administration does not claim that its targeted killing policies, as opposed to some other aspects of the conflict, are an exercise of its law enforcement powers. Targeted killing is war, the administration asserts, and for good reason. "Since many US actions using lethal force [including target killings] would constitute murder or other crimes during peacetime, establishing the legal propriety of a state of armed conflict is actually an important point."[55] In order to evaluate the constitutionality of targeted killing, therefore, we need to consider the framework of war powers that the Constitution delegates to the government of the United States.

54. On the administration's position, *see id.* at 22–25. Administration speeches "consistently emphasize law enforcement as the chief—even exclusive—tool for use at home and against Americans *captured* anywhere in the conflict." *Id.* at 24. The *Hamdi v. Rumsfeld* line of decisions makes it clear that the executive is constitutionally obligated to adopt the law enforcement paradigm in detaining, in the United States or at Guantanamo, citizens it believes to enemy combatants who dispute that characterization. *Boumediene v. Bush*, which held that alien detainees at Guantanamo are entitled to the privilege of the writ of habeas corpus absent a valid congressional suspension of the writ, did not reach the question whether "the President has authority to detain these petitioners." 553 U.S. 723, 733 (2008). *Boumediene* illustrates the fact that even when the government exercises its powers to wage war it does so subject to the constraints imposed by "fundamental separation-of-powers principles." *Id.* at 755. *See also id.* at 765 ("the writ of habeas corpus is itself an indispensable mechanism for monitoring the separation of powers").

55. ANDERSON & WITTES, *supra* note 53, at 22. Anderson and Wittes have in view the criminal laws of other countries as well as the United States, but their observation is clearly the case with respect to U.S. law and the targeted killing of U.S. citizens. *See* 18 U.S.C. § 1119(b) (making it a federal crime to kill unlawfully a U.S. national abroad).

Chapter 3

The War Powers of the U.S. Government

In deciding whether it was constitutional for the U.S. government to kill Anwar al-Awlaki by drone strike, we must apply the tools of sound constitutional thinking (the subject of the previous chapter) within the specific framework of analysis that constitutional text and tradition have established. As this chapter will show, the U.S. Constitution's arrangements for war are not simple: The federal government's war powers are plenary, limited, and divided.

A. FEDERAL WAR POWERS ARE PLENARY

A preliminary point may be important: There is no conflict between asserting that a federal power is plenary and at the same time that the power is limited. In the famous 1824 case of *Gibbons v. Ogden*, Chief Justice Marshall described Congress's powers over international, interstate, and Indian commerce as "plenary," meaning by the word that the Constitution authorizes Congress to exercise full regulatory authority over international, interstate, and Indian commerce. "This power, like all others vested in Congress, is complete in itself, may be exercised to its utmost extent, and acknowledges no limitations, other than are prescribed in the constitution."[1] But Marshall did not mean by this that the "plenary" and "complete" commerce power is conceptually unlimited: It is, after all, authorization to exercise a certain power (or powers) over a specific set of objects (however broadly defined) rather than a grant of *all* legitimate legislative powers. "The enumeration [of some objects of regulation] presupposes something not enumerated," a principle that in *Gibbons* Marshall specifically applied to differentiate the "commerce" Congress can regulate from that which it cannot. Constitutional authorizations confer the authority to do some things and not others; all federal powers are limited by the implicit definitions of their scope.

All of these general observations about the powers of the federal government apply fully to the powers that relate to war. In evaluating the validity of a federal action predicated on the war powers, we should assume those powers are "complete in [themselves]," extending to the full scope of the underlying objects summarized by the word "war." But not everything is "war," or encompassed in the various clauses relating to "war," and so the invocation of "war" as a justification does not automatically prove a federal action constitutional.[2] Furthermore, sound constitutional thinking will respect the distinctive nature of the various war power clauses. However broadly we should construe

1. 22 U.S. (9 Wheat.) 1, 196–97 (1824).

2. The principle of enumeration is not the only limitation on federal power. In addition, as discussed in the previous chapter, there are constitutional prohibitions to which all authorizations of federal power are subject. Furthermore, as the Supreme Court reaffirmed in *Boumediene v. Bush*, the checks and limitations on congressional and executive authority implicit in "the Constitution's separation-of-powers structure" remain effective and, if necessary, judicially enforceable even in a time of war. "The laws and Constitution are designed to survive, and remain in force, in extraordinary times." 553 U.S. 723, 797, 798 (2008).

its scope, for example, the Marque and Reprisal Clause of Article I, section 8 has no obvious relevance to the determining the constitutionality of targeted killing.[3]

In the U.S. constitutional system, then, the language of "plenary" power is not synonymous with "unlimited" power. With this in mind, we can examine the affirmative implications of recognizing the plenary nature of the federal powers over war-related matters. Their completeness, in Chief Justice Marshall's terms, has been clear from before the beginning of our current constitutional order. In *Federalist* No. 23, Alexander Hamilton wrote:

> The authorities essential to the common defense are these: to raise armies; to build and equip fleets; to prescribe rules for the government of both; to direct their operations; to provide for their support. These powers ought to exist without limitation: *Because it is impossible to foresee or define the extent and variety of national exigencies, or the correspondent extent and variety of the means which may be necessary to satisfy them.* The circumstances that endanger the safety of nations are infinite, and for this reason no constitutional shackles can wisely be imposed on the power to which the care of it is committed. This power ought to be coextensive with all the possible combinations of such circumstances; and ought to be under the direction of the same councils which are appointed to preside over the common defense.

3. *See* U.S. CONST. art. I, § 8, cl. 11 (delegating to Congress the power "to grant Letters of Marque and Reprisal"). In eighteenth-century law, a letter of marque and reprisal authorized a private vessel to capture enemy vessels, i.e., it was the governmental license to act as a privateer.

James Madison, usually thought less hawkish than Hamilton, wrote to identical effect in *Federalist* No. 41.[4] That Hamilton and Madison not only admitted but insisted on the plenary nature of the war powers the Constitution would confer is all the more remarkable when one recalls that the anti-federalist opponents of ratification made much out of what they saw as the dangerously broad grant of war powers being proposed. Hamilton and Madison, and other prominent advocates of ratification, seldom missed the opportunity to rebut anti-federalist alarms by describing the powers of the new federal government in muted, if not minimizing, language, but as to war they apparently thought it neither prudent nor principled to deny the plenary nature of the authority the proposed Constitution would grant the federal government over war.

Over a century later, shortly after the United States entered the First World War, another distinguished constitutionalist, Charles Evans Hughes, gave an address to the American Bar Association on "War Powers Under the Constitution." Hughes, a former Associate Justice and later to be Chief Justice of the United States, insisted that the transition to war changed nothing about the structure and limited nature of federal power. "While we are at war, we are not in revolution. We are making war as a Nation organized under the constitution, from which the established national authorities derive all their powers either

4. THE FEDERALIST No. 23 (Alexander Hamilton), at 147 (Jacob E. Cooke ed., 1961); *id.* No. 41 (James Madison), at 269–70 ("Security against foreign danger is one of the primitive objects of civil society. It is an avowed and essential object of the American Union. The powers requisite for attaining it must be effectually confided to the federal councils. . . . With what color of propriety could the force necessary for defense be limited by those who cannot limit the force of offense? If a federal Constitution could chain the ambition or set bounds to the exertions of all other nations, then indeed might it prudently chain the discretion of its own government, and set bounds to the exertions for its own safety.").

in war or in peace. The constitution is as effective today as it ever was." But Hughes went on to echo the *Federalist*: "The power to wage war is the power to wage war successfully. . . . In equipping the National Government with the needed authority in war, [the founders] tolerated no limitations inconsistent with that object."[5]

In *Lichter v. United States*, the Supreme Court quoted Hamilton, Madison, and Hughes in upholding what the Court described as "a new legislative solution" to addressing "the problem of national defense against world-wide aggression" in "unprecedented" circumstances.[6] The Court borrowed words from Hughes's address in underlining its insistence that the government's war powers are adequate to address any foreign military threat no matter how novel: "'[T]here are constantly new applications of unchanged powers, and . . . in novel and complex situations, the old grants contain, in their general words and true significance, needed and adequate authority.'"

> [I]n the light of the language just quoted, the primary implication of a war power is that it shall be an effective power to wage the war successfully. Thus, while the constitutional structure and controls of our Government are our guides equally in war and in peace, they must be read with the realistic purposes of the entire instrument fully in mind.[7]

The Court's reasoning in *Lichter* followed the approach to constitutional thinking outlined by Chief Justice Marshall in *McCulloch v. Maryland*: We should read the words of the constitutional text insofar as is possible in light of the purposes of

5. Hughes, *War Powers Under the Constitution*, 2 MARQUETTE L. REV. 3, 3, 9 (1917).

6. 334 U.S. 742, 745 (1948).

7. *Id.* at 781–82.

the American constitutional enterprise. Nothing has happened since *Lichter* that puts into question the view of the war powers on which *Lichter* rested: The war powers are plenary, in the particular sense that term possesses in U.S. constitutional law. The Constitution delegated to the political branches various specific powers relating to military matters and war, and did so to the end that the federal government should "wage war successfully" in response to whatever dangers to the security of the Republic the future might bring, regardless of how unforeseeable or undefinable those dangers were at the time of the founding.

Musketeers in uniform and ships of the line under sail were a foreseeable threat to national security at the end of the eighteenth century; terrorists hijacking airliners turned out to be an unforeseen military threat at the beginning of the twenty-first. From the standpoint of the constitutional authorizations of power, what the threat is does not matter: The answer in either and any case is that the federal government is empowered to take the steps it deems necessary to preclude or defeat the military threat. "The constitutional question presented in the light of an emergency is whether the power possessed embraces the particular exercise of it in response to particular conditions. Thus, the war power of the federal government is not created by the emergency of war, but it is a power given to meet that emergency. It is a power to wage war successfully."[8] In the light of constitutional purpose, the reasoning in Supreme Court precedent, and political practice, there is no serious doubt about the proposition that the Constitution provides the government power equal to every

8. Home Bldg. & Loan Ass'n v. Blaisdell, 290 U.S. 398, 426 (1934) (Hughes, C.J., for the Court). Chief Justice Hughes felt free to use the language of his ABA address without formally quoting himself.

possible national security emergency, including the threat posed by contemporary terrorism.[9]

The plenary nature of the federal war powers answers, as a matter of U.S. constitutional law, an important question that many critics have raised about the executive branch's use of drone strikes as a method of targeted killing. The issue stems from the fact, as it seems to be, that the personnel carrying out many drone strikes are CIA employees or agents rather than members of the U.S. armed forces.[10] The legality of this practice under international law is unclear.[11] In contrast, I think there is little room to doubt that it is lawful for Congress to authorize the president to use the CIA to carry out lethal actions in the war on terror. There is clearly no absolute constitutional bar on the United States

9. I am paraphrasing the language of a Virginia state court case upholding congressional power to institute a military draft cited by the Supreme Court in *Lichter. See* 334 U.S. at 756 n.4 (quoting Burroughs v. Peyton, 57 Va. (16 Gratt.) 470 (1864)). *Lichter* quoted language from the *Burroughs* opinion coming shortly before the passage I have in mind: "A nation cannot foresee the dangers to which it may be exposed: it must therefore grant to its government a power equal to every possible emergency." 57 Va. (16 Gratt.) at 473. The Virginia court supported this general reasoning with the observation that "one of the chief reasons urged in favor of the change of the form of government effected by the adoption of the constitution of the United States" was the founders' insistence that the federal government "charged with the duty of providing for the common defence should be invested with power commensurate with that end." *Id.* at 474. I do not know if Justice Harold Burton, who wrote for the Court in *Lichter*, consciously intended the irony implicit in relying on a decision upholding the power of the *Confederate* States government to wage war successfully—and doing so immediately after quoting President Lincoln! In any event, *Lichter* and *Burroughs* both express what I believe is the clearly correct view of the scope of the war powers of the U.S. government.

10. *See, e.g.*, Deep Support in Washington for C.I.A.'s Drone Missions (Apr. 25, 2015) (discussing CIA's role in drone strikes as of spring 2015), http://www.nytimes.com/2015/04/26/us/politics/deep-support-in-washington-for-cias-drone-missions.html?_r=0; Philip Alston, *The CIA and Targeted Killings Beyond Borders*, 2 HARV. NAT'L SEC. J. 283, 328–33 (2011) (discussing earlier history of CIA involvement).

11. *See, e.g.*, Donna R. Cline, *An Analysis of the Legal Status of CIA Officers Involved in Drone Strikes*, 12 SAN DIEGO INT'L L.J. 51 (2013).

using persons in military operations who are not members of the regular armed forces: In addition to providing for the use of state militia "in the Service of the United States," Article I expressly authorizes Congress to issue letters of marque and reprisal and thereby license private persons to act as privateers. The *McCulloch* principle of broad congressional discretion, and long-standing political practice as well, support the constitutional validity of Congress structuring the public organizations that carry out military and quasi-military operations without regard to the constitutional text's literal description of the forces available to the commander in chief as "the Army and Navy [and] Militia [when in] the actual Service of the United States."[12] It would be absurd to argue that the changing institutional histories of the Air Force, the Marine Corps, the Coast Guard, and the National Guard are constitutionally problematic, and Congress's authorization of covert action by the CIA demonstrates the legislature's judgment that that agency too can be employed in national security operations that may involve the use of military force.[13] Whether the CIA has the lawful authority to carry out a particular drone strike depends entirely on the executive's compliance with the statutes governing CIA activities and raises no constitutional issues.

B. FEDERAL WAR POWERS ARE LIMITED

The war powers of the federal government are plenary, but they are also limited, not only conceptually as we just saw (they address military issues rather than, say, intellectual property) but also

12. U.S. CONST. art. II, § 2.

13. *See* 50 U.S.C.A. § 3093 (detailing required criteria and procedures for presidential authorization of "covert action").

in that they are subject to override by applicable constitutional prohibitions. This general principle applies to the war powers as much as to those concerning domestic governance. No matter how "plenary" the war powers of the president or Congress—or of the president and Congress acting together—they cannot be exercised lawfully in a manner that violates an applicable prohibition such as the First Amendment.[14] Much of the debate over targeted killing since the death of al-Awlaki has revolved around the question whether al-Awlaki's killing, or the administration's policy more generally, violates the Fifth Amendment, which provides that no "person" shall "be deprived of life, liberty, or property, without due process of law." The amendment does not use the term "citizen," and when applied to the federal government's domestic actions the Due Process Clause is not restricted to U.S. citizens, but most discussions of due process and targeted killing have assumed that the clause does not protect noncitizens outside the jurisdiction of the United States.[15] The critics have vigorously argued that the administration's policy patently fails to provide citizen-targets with due, or indeed any, process; the administration's lawyers have assumed (and the president himself has asserted) that due process is owed to a citizen-target such as Anwar al-Awlaki. On that assumption, the administration's position is that the procedures required under its current policies satisfy the Fifth Amendment's requirements.

14. *See, e.g.*, New York Times v. United States, 403 U.S. 713 (1971) (holding that the first amendment prohibition on prior restraints forbade the issuance of an injunction to prevent the disclosure of information although the executive, and some of the Justices, believed disclosure would do serious harm to the foreign policy and national defense interests of the United States).

15. *See, e.g.*, Zadvydas v. Davis, 533 U.S. 678, 693 (2001) ("It is well established that certain constitutional protections available to persons inside the United States are unavailable to aliens outside of our geographic borders [including the] Fifth Amendment's protections").

What critics and apologists alike have failed to consider is the possibility that the Due Process Clause simply doesn't apply to the targeted killing of a citizen-enemy such as al-Awlaki. Let us consider that possibility.

Begin with a broad and uncontroversial proposition: Constitutional prohibitions resemble constitutional authorizations in that they are conceptually limited. Even the most robust reading of the Ex Post Facto Clause, for example, cannot make its prohibition relevant to an executive branch action such as the executive's decision to kill al-Awlaki.[16] Like all other constitutional prohibitions, therefore, the Fifth Amendment Due Process Clause controls certain exercises of federal power but not others. A familiar example in the case law concerns the right of aliens to procedural due process. The Supreme Court "has rejected the claim that aliens are entitled to Fifth Amendment rights outside the sovereign territory of the United States," but "once an alien enters the country, the legal circumstance changes, for the Due Process Clause applies to all 'persons' within the United States, including aliens."[17] There is no obvious reason why this distinction should apply only to the Fifth Amendment rather than extending to the other prohibitions contained in the Bill of Rights, and a 1990 decision, *United States v. Verdugo-Urquidez*,

16. U.S. CONST. art. I, § 9, cl. 23 ("No . . . ex post facto Law shall be passed."). "The Ex Post Facto Clause is a limitation upon the powers of the Legislature." Marks v. United States, 430 U.S. 188, 191 (1977).

17. United States v. Verdugo-Urquidez, 494 U.S. 259, 269 (1990); Zadvydas v. Davis, 533 U.S. 678, 693 (2001). The U.S. Naval Station at Guantanamo may be an exception to this principle as it is in other respects because of its sui generis status: Its territory is theoretically under the ultimate sovereignty of Cuba, but in fact and law subject to plenary U.S. control. *See, e.g.*, Boumediene v. Bush, 553 U.S. 723, 769 (2008) ("In every practical sense Guantanamo is not abroad; it is within the constant jurisdiction of the United States").

held that that the principle applies to the Fourth Amendment prohibition on unreasonable searches and seizures as well.[18]

Under aliens-abroad decisions such as *Verdugo-Urquidez*, drone strikes in foreign territory against noncitizens raise no constitutional issues under the Fourth and Fifth Amendments, but Anwar al-Awlaki *was* a U.S. citizen, and the law could be what critics assume it is, that the government must satisfy the requirements of the Due Process Clause when an American is the target. However, under a long line of Supreme Court decisions, it is in fact clear that for reasons other than geography, Bill of Rights prohibitions do not apply to legitimate military exercises of the federal government's war powers. The conceptual basis for this rule lies in the distinction the Court drew at a very early point between the exercise of domestic legislative authority over those being governed and the exercise of war powers against an enemy who is entirely outside the political community. The most extensive discussions are found in cases arising out of the Civil War, although the earliest decisions in the series date from the tenure of Chief Justice Marshall and the most recent, Ex parte *Quirin*, from World War II.[19]

18. The Supreme Court relied on the Fifth Amendment decisions about aliens to support its holding that the Fourth Amendment is equally inapplicable to aliens outside the United States. *Verdugo-Urquidez*, 494 U.S. at 269. The Court also observed more generally that in the founding era "it was never suggested that the Fourth Amendment restrained the authority of Congress or of United States agents to conduct [naval] operations." *Id.* at 268 (discussing operations by U.S. warships and privateers during the 1798-1800 undeclared naval war with France).

19. There is no need to review each of the decisions, although it may be worth noting how early certain key principles emerged. In *Rose v. Himely*, 8 U.S. (4 Cranch) 241 (1808), *overruled on other grounds*, *Hudson v. Guestier*, 10 U.S. (6 Cranch) 281 (1810), the Constitution was not at issue, but Chief Justice Marshall's opinion introduced two principles that the Court subsequently found relevant to constitutional thinking. First, Marshall distinguished the "sovereign power" to enact "municipal regulations" from a nation's "belligerent rights" (war powers), and suggested that war powers cannot lawfully be employed "to execute a municipal law." *Id.* at 272–73, 279. The exercise of war powers makes available means that are not lawful when used in executing ordinary domestic legislation. Second, in concluding that

Among these cases, the Civil War decisions are particularly significant for our purpose because Congress and President Lincoln were clearly exercising the federal war powers against an enemy made up almost entirely of U.S. citizens: The legal foundation of the government's use of military force, subsequently vindicated by the Supreme Court, was that secession was unconstitutional and that the United States was engaged not in a war of

the federal courts should not enforce a French judgment because the French court was enforcing French domestic law rather than acting as a prize court in wartime, Marshall reasoned that in a civil war, a national government may treat rebels in a given case either as criminals subject to its ordinary domestic law, or as enemies with whom it is engaged in a "war *de facto*," but it cannot do so in one and the same act. *See id.* at 272 ("But admitting a sovereign who is endeavouring to reduce his revolted subjects to obedience, to possess both sovereign and belligerent rights, and to be capable of acting in either character, the manner in which he acts must determine the character of the act.").

In *Brown v. United States*,12 U.S. (8 Cranch) 110 (1814), the Court held that Congress's declaration of war on Great Britain did not, of its own force, confiscate timber in the United States belonging to a British subject who had subsequently sold the property to a U.S. citizen. In his opinion for the Court, Chief Justice Marshall asserted that "[r]especting the power of government no doubt is entertained. That war gives to the sovereign full right to take the persons and confiscate the property of the enemy wherever found, is conceded, is conceded. The mitigations of this rigid rule, which the humane and wise policy of modern times has introduced into practice, will more or less affect the exercise of this right, but cannot impair the right itself. That remains undiminished." *Id.* at 122–23. In part because the Constitution was written after the emergence of the "humane and wise policy" of permitting alien enemies the opportunity to remove their property at the onset of war, Marshall thought the Declaration of War Clause ought to be construed to authorize Congress to exercise its discretion on the matter: The force of the declaration in itself was to render lawful the exercise of war powers such as the seizure of persons and private property: "[T]he declaration of war has only the effect of placing the two nations in a state of hostility, of producing a state of war, of giving those rights which war confers." *Id.* at 125–26. If Congress had chosen to do so, its war powers would have permitted it to authorize the confiscation of the private property at issue.

Many of the mid- and late-nineteenth-century cases, including those arising out of the Civil War, are summarized in *Juragua Iron Co. v. United States*, 212 U.S. 297 (1909), which rejected a claim for just compensation by an American company with extensive real property holdings in Cuba that were deliberately destroyed by U.S. armed forces during the Spanish-American War. The Court conceded that the American troops were under "a duty to observe the rules governing the conduct of independent nations when engaged in war,—a duty for the proper performance of

aggression aimed at the conquest of an independent nation (the Confederate view, of course) but in the suppression of a massive insurrection.[20]

This view of the war vindicated the moral and political legitimacy of the Union war effort, but at the cost of raising a serious constitutional difficulty. The secessionists' actions in taking over federal property and interfering with federal governmental functions were clearly criminal; no later than the attack on Fort Sumter, and arguably earlier than that, secession had become

which the United States may have been responsible in its political capacity to the enemy government," but "the plaintiff corporation could not invoke the protection of the Constitution in respect of its property used in business in Cuba, during the war, any more than a Spaniard residing there could have done, under like circumstances." *Id.* at 308.

The World War II decision, Ex parte *Quirin*, 317 U.S. 1 (1942), upheld the lawfulness of trying by military commission several German saboteurs who were apprehended within the United States and charged with violations of the laws of war. One of the saboteurs, Herbert Haupt, was a citizen, but the Court dismissed this as irrelevant: "[E]ven when committed by a citizen, the [law of war] offense is distinct from the crime of treason defined in Article III, § 3 of the Constitution." "Citizenship in the United States of an enemy belligerent does not relieve him from the consequences of a belligerency which is unlawful because in violation of the law of war." *Id.* at 38, 37. As an enemy belligerent, Haupt stood in a fundamentally different relationship to the U.S. government than would a disloyal U.S. citizen arrested on U.S. soil for traitorous activity but who was "not . . . a part of or associated with the armed forces of the enemy, was a non-belligerent, not subject to the law of war save as . . . martial law might be constitutionally established." *Id.* at 45.

For detailed discussions of this body of law, the reader should turn to two outstanding articles from which I have learned a great deal. *See* John C. Dehn, *The Commander in Chief and the Necessities of War: A Conceptual Framework*, 83 Temp. L. Rev. 599, 621–43 (2011); Andrew Kent, *The Constitution and the Laws of War During the Civil War*, 85 Notre Dame L. Rev. 1839, 1893–1927 (2010).

20. "Considered therefore as transactions under the Constitution, the ordinance of secession, adopted by the convention and ratified by a majority of the citizens of Texas, and all the acts of her legislature intended to give effect to that ordinance, were absolutely null. . . . It certainly follows that the State did not cease to be a State, nor her citizens to be citizens of the Union. If this were otherwise, the State must have become foreign, and her citizens foreigners. The war must have ceased to be a war for the suppression of rebellion, and must have become a war for conquest and subjugation." Texas v. White, 74 U.S. 700, 726 (1868).

treason within the constitutional definition—"levying War against" the United States, "or . . . adhering to their Enemies, giving them Aid and Comfort."[21] But crimes, including the high crime of treason, are matters for arrest and criminal prosecution. To be sure, Article I authorizes Congress to empower the president to "call[] forth the Militia to execute the Laws of the Union [and] suppress Insurrections," but this use of military force is no more than a means for enforcing the laws against criminal "'combinations too powerful to be suppressed by the ordinary course of judicial proceedings, or by the power vested in the marshals.'" In such an extraordinary situation, the use of military force is not to make war, and it "can do no more than what might and ought to be done by a civil *posse*, if a civil *posse* could be raised large enough to meet the same opposition."[22]

Near the end of the Buchanan administration, Attorney General Jeremiah S. Black wrote a formal opinion arguing that the congressional authority to declare war is limited to "the power to commence and carry on hostilities against the foreign enemies of the nation," and that in a state where the federal government had ceased to function, the president's statutory power to use military force to execute the laws would be nugatory: "Under such circumstances, to send a military force into any State, with orders to act against the people, would be simply making war upon them."[23] In the *Prize Cases*, the Supreme Court agreed with Attorney General Black that Congress and the president have no constitutional authority to *choose* to employ the war powers against U.S. citizens in lieu of exercising ordinary powers of

21. U.S. CONST. art. III, § 3.
22. *Power of the President in Executing the Laws*, 9 Op. Att'y Gen. 516, 522 (1860).
23. *Id.* at 524, 523.

governance.[24] Nothing about the Court's decision in that or any of the other Civil War cases suggests that the political branches can proceed against a U.S. national as an enemy pursuant to the war powers rather than as a citizen subject to civil regulation and criminal law because as a policy matter they would prefer to do so. President Lincoln's exercise of his military powers as commander in chief, and Congress's subsequent ratification and extension of his war measures, were *not* constitutionally authorized because it was more convenient to fight the Confederate army rather than arrest its members as traitors.

Nonetheless, the Court reasoned, the executive and Congress were acting constitutionally in employing the war powers against Confederates as enemies. The constitutional authorization for the use of the war powers against these U.S. citizens and the rebellious state governments rested on the Constitution's realistic anticipation of an armed rebellion that forces what is in military fact a war on the federal government. Such a rebellion poses a military threat to the Republic itself, one that as a practical matter cannot be dealt with through the modes of legal process available to the powers of governance; as the Court later asserted: "[I]t would savor somewhat of the puerile and ridiculous to have read a writ of injunction to Lee's army during the late Civil War."[25] The nineteenth Court's logic on this issue was sound. If we approach constitutional issues as *McCulloch* teaches, it is clear that in a circumstance where only the war powers are adequate to protect the Republic, the Constitution authorizes the political branch to employ those powers, even if U.S. citizens are the source of military threat to the Republic.

24. 67 U.S. 635 (1863).
25. In re *Debs*, 158 U.S. 564, 597 (1895).

Where U.S. citizens are involved in an armed conflict against the Republic and cannot be dealt with as ordinary criminals, then, the political branches may make war on them. From this conclusion, the Civil War cases drew another. No one would suggest that the soldiers, or more generally the citizens, of a foreign state with which the United States was engaged in war under the law of nations would be protected by the constitutional prohibitions that constrain the federal government's powers of domestic governance; in such a war, it would be obvious that as Chief Justice Marshall put it in 1808, the federal government was exercising its "belligerent rights" rather than its "sovereign power" to enact "domestic regulations."[26] "War existing, the United States [is] invested with belligerent rights in addition to the sovereign powers previously held."[27] In making war, the federal government is not governing the community but defending it, and this fundamental shift in constitutional role is reflected in sound constitutional thinking. The Bill of Rights imposes limitations on domestic governance as part of the fundamental constitutional commitments to limited government and the protection of liberty, but those commitments are not in question when the government acts against those whom it cannot as a practical matter govern. The relevant constitutional commitment in that circumstance is to the principle that the federal government has the power to wage war successfully. When the political branches are constitutionally authorized to direct the war powers against a citizen as an enemy, therefore, they are not subject to Bill of Rights constraints.

The Supreme Court repeatedly acted on this reasoning in connection with the Civil War. In *Miller v. United States*, for example,

26. *Rose*, 8 U.S. (4 Cranch) at 272–73, 279.
27. Miller v. United States, 78 U.S. 268, 306 (1870).

the Court upheld the application of wartime confiscation laws to railroad stock belonging to a Virginia citizen who had adhered to the Confederacy. The Court rejected the argument that the statutes' summary process and substantive destruction of the individual's property rights were violations of the Fifth and Sixth Amendments. Justice Strong conceded that "if the [laws] were municipal regulations only, there would be force in the objection that Congress has disregarded the restrictions of the fifth and sixth amendments of the Constitution."

> But if . . . on the contrary, [the laws] are an exercise of the war powers of the government, it is clear they are not affected by the restrictions imposed by the fifth and sixth amendments. . . . Of course the power to declare war involves the power to prosecute it by all means and in any manner in which war may be legitimately prosecuted. It therefore includes the right to seize and confiscate all property of an enemy and to dispose of it at the will of the captor. . . . The whole doctrine of confiscation is built upon the foundation that it is an instrument of coercion [i.e., and not a means of punishing the individual].[28]

A use of military force that is lawful as a matter of constitutional authorization, therefore, need not conform to Bill of Rights prohibitions even when the military force is directed against U.S. citizens.

Under the logic of these cases, in the context of a constitutionally authorized war, a citizen-enemy against whom the

28. 78 U.S. at 306. The Court's reference to "any manner in which war may be *legitimately* exercised" was an allusion to constraints on war making derived from the law of nations: The Civil War cases consistently assumed that those constraints were of normative but extraconstitutional significance.

government may act militarily remains liable as a citizen to government actions based on the ordinary legislative powers of governance. The actions that under Article II constitute treason are belligerent acts against which the government may exercise its war powers, but since they are also a crime, the government may prosecute the traitor through the ordinary methods of law enforcement.[29] It would have been constitutionally permissible to prosecute a captured Confederate soldier for treason using the powers of governance. If the government had decided to do so, the soldier would have been entitled to all the criminal procedure guarantees of the Bill of Rights[30] and he could not have been executed or otherwise punished for the crime until a federal court lawfully convicted him. On the other hand, exercising its war powers, the government was constitutionally authorized to kill the Confederate soldier on the battlefield without regard to the Bill of Rights, or if it captured him, it could detain him as a prisoner of war until the end of the conflict, again without the obligation to respect such constitutional prohibitions as the Due Process Clause. *Miller* is representative of this long line of decisions that rest on the principle that the Constitution's prohibitions simply do not apply to the exercise of the war powers against an enemy even though a citizen who is a wartime enemy

29. The district judge in *The Amy Warwick*, one of the cases that were consolidated in the Supreme Court as the *Prize Cases*, summarized the view taken in all of the Civil War decisions: Citizen enemies "are at the same time belligerents and traitors, and subject to the liabilities of both; while the United States sustains the double character of a belligerent and sovereign, and has the rights of both. These rights co-exist, and may be exercised at pleasure. Thus, we may treat the crew of a rebel privateer merely as prisoners of war, or as pirates, or traitors; or we may, at the same time, give to a part of the crew the one character, and to the residue the other." *The Amy Warwick*, 1 Fed. Cas. 799, 803 (D. Mass. 1862). *See also* Kent, *supra* note 19, at 1839, 1884–85, 1918–19.

30. A Confederate soldier/treason defendant would also have been entitled to the evidentiary requirement of Article III, section 3, clause 1 and the substantive restrictions on criminal punishment in Article III, section 3, clause 2 and the Eighth Amendment.

remains subject to domestic legal processes (with their accompanying constitutional limitations) as well.[31]

On the face of it, cases like *Miller* provide a clear response to the charge that Anwar al-Awlaki was deprived of his life without due process: according to those cases, al-Awlaki was entitled to due process in the context of a military operation against him as an enemy combatant. The main objection to treating the old cases as controlling is the widespread intuition, memorably expressed by President Obama, that it is unthinkable that the U.S. government can kill U.S. citizens without first affording them due process. In itself, that intuition reflects a healthy sense of the government's subordination to law; furthermore, constitutional law need not, and indeed I think should not, be walled off from popular intuitions about justice, but equally such intuitions are not in themselves a constitutional argument.[32] Most participants in the current debate do not seem to find the killing of noncitizens without due process to be unthinkable, so the issue is not one of barbaric versus civilized acts of war: If targeted killing is

31. *See* Dehn, *supra* note 19, at 641 (the nineteenth-century cases "support[] the general proposition, that those properly considered public enemies, whether citizens or foreign nationals, may not claim the protections of the Bill of Rights against a proper use of war or military occupation powers.").

32. A well-known line from Chief Justice Hughes's opinion for the Court in the *Blaisdell* decision does give me pause: "[E]ven the war power does not remove constitutional limitations safeguarding essential liberties." Home Building & Loan Assn. v. Blaisdell, 290 U.S. 398, 426 (1934). The footnote Hughes appended to this statement refers to a well-known case, Ex parte *Milligan*, 71 U.S. (4 Wall.) 2 (1866), in which the war powers were being directed against someone thought by the executive to be an enemy. But the decision in *Milligan*, as Justice O'Connor pointed out later in *Hamdi v. Rumsfeld*, "turned in large part on the fact that Milligan was not a prisoner of war but a resident of Indiana arrested while at home there."

> That fact was central to its conclusion. Had Milligan been captured while he was assisting Confederate soldiers by carrying a rifle against Union troops on a Confederate battlefield, the holding of the Court might well have been [that] he could have been detained under military authority for the duration of the conflict, whether or not he was a citizen.

horrific, then it is horrific regardless of the victims' identity. To say that the Due Process Clause does not prohibit the executive's use of drone strikes against (citizen) enemy combatants in the current authorized conflict does not imply that the executive can employ targeted killing as a law enforcement measure: The *Prize Cases*, and the logic of the *Miller* line of decisions, rule out the use of the war powers as an ordinary tool of governance.[33]

Supreme Court decisions do not come with sell dates or fall into a state of decisional desuetude: The mere fact that the nineteenth-century cases are old does not demonstrate that they are no longer good law.[34] Furthermore, the argument that they were correctly decided on principle is firmly rooted in the *McCulloch* perspective on federal power. The constitutional prohibitions imposed by the Bill of Rights and elsewhere safeguard individual liberty against the exercise of power to govern the political community, but the

Hamdi, 542 U.S. 507, 522 (2004) (plurality opinion). O'Connor also quoted Hughes's dictum, but she did so in support of her rejection of the executive's argument that "separation of powers principles mandate a heavily circumscribed role for the courts" in detention decisions—in other words her primary focus at that point in her opinion was on the liberty-protecting purpose of separation of powers. *See id.* at 535. All of this said, it makes me as uncomfortable as it likely does the reader to follow the logic of the nineteenth-century cases, and Chief Justice Hughes's memorable words encapsulate part of the discomfort. But sound constitutional thinking is not a panacea for all societal ills, and a constitutional conclusion that seems correct but disturbing is likely to be an indication that we should view the nonconstitutional considerations as critical.

33. The extent to which Congress may authorize the executive to employ the armed forces to carry out law enforcement duties, at least in the absence of a complete breakdown of the ordinary criminal justice system, is unclear and need not be resolved for our purposes. Military personnel doing so pursuant to valid legislative authority would be subject to the constitutional prohibitions that constrain civilian law enforcement officials. *Cf.* Duncan v. Kahanamoku, 327 U.S. 304, 322-23 (1946) ("Our system of government clearly is the antithesis of total military rule ... 'the military should always be kept in subjection to the laws'").

34. As noted above, there are twentieth-century decisions that carry forward the principle. *See* Dehn, *supra* 19, at 643 (the World War II decision in Ex parte *Quirin* "implicitly [held] that only the laws of war govern the rights of enemies in armed conflict with the United States").

war powers safeguard the community itself against threats that potentially may imperil its very survival. Because Congress and the president cannot elect at their discretion to substitute their war powers for the ordinary powers of governance within the political community, the danger the lawful use of military force poses to liberty is tightly confined by the limits on the purposes for which the war powers may legitimately be used.

The Supreme Court's recent decisions regarding the post-9/11 detention of captured enemy combatants do not substantially modify the principle followed in the nineteenth-century cases and more recently in Ex parte *Quirin*. The later of the two major decisions, *Boumediene v. Bush* did not address the constitutional rights of U.S. citizens or indeed any question about the substantive rights of anyone: It was fundamentally a separation of powers decision holding that a noncitizen detained within the jurisdiction of the United States is entitled to test the legality of his detention by invoking the habeas corpus jurisdiction of the federal courts unless Congress has validly suspended the privilege of the writ.[35] *Boumediene* in fact indicates the basis for distinguishing the Suspension Clause, which the Court treated as primarily a safeguard for the Constitution's dispersal of federal power among the three branches, from the Bill of Rights constitutional prohibitions that limit the exercise of power by all three branches of the federal government.[36]

35. The *Boumediene* Court repeatedly invoked separation of powers principles and characterized its decision as resting on the principle that "personal liberty that is secured by adherence to the separation of powers." 553 U.S. at 797. The Court expressly declined to reach any substantive issues. *See id.* at 798 ("It bears repeating that our opinion does not address the content of the law that governs petitioners' detention. That is a matter yet to be determined.").

36. *See id.* at 743, 745 (habeas corpus is "an essential mechanism in the separation-of-powers scheme" which ensures that "except during periods of formal suspension, the Judiciary will have a time-tested device, the writ, to maintain the 'delicate balance of governance' that is itself the surest safeguard of liberty").

The other major detention case, *Hamdi v. Rumsfeld*, involved a U.S. citizen whom the executive was detaining as an enemy combatant pursuant to the Authorization for the Use of Military Force but who insisted that he was a bystander caught up in the Afghanistan war. The controlling opinion by Justice Sandra Day O'Connor concluded that a U.S. citizen detained on U.S. soil "as an enemy combatant" is entitled to Fifth Amendment procedural due process "to contest the factual basis for that detention." The holding in *Hamdi* is entirely consistent with the nineteenth-century cases involving the military seizure or destruction of private property in enemy territory or owned by citizens whose status as enemies was established. Justice O'Connor's opinion is equally consistent with Civil War practice. During the conflict, the federal government detained captured Confederate soldiers without prosecuting them or allowing them to contest their indefinite detention. No process was afforded them, but it was clear, from concession or circumstance, that all or virtually all of these U.S. citizen detainees were enemy combatants, while the very question at issue in *Hamdi* was whether that citizen could in fact be treated lawfully as an enemy and subjected to military action as such. O'Connor's reasoning allows the treatment of a citizen as an enemy pursuant to the lawful exercise of the war powers without regard to due process or other procedural guarantees in the Bill of Rights "once it is sufficiently clear that the individual is, in fact, an enemy combatant."[37]

If this is the right way to read *Hamdi*, then the decision casts no doubt on the continuing validity of the line of cases including *Miller*. Justice O'Connor did not claim to overrule or even

37. 542 U.S. at 509. *See also* her discussion of Ex parte *Milligan* discussed above. In contrast, Justice Antonin Scalia (joined by Justice John Paul Stevens) would have required the executive to release Hamdi or prosecute him for a crime: In other words, Scalia disagreed with the principle established in the nineteenth-century cases, of which he seems to have been unaware. *See id.* at 573 (Scalia, J., dissenting).

limit any of the *Miller* line of decisions, and she treated Ex parte *Quirin* (which is to the same effect) as precedent. If *Hamdi* should be treated as overruling the earlier cases *sub silentio*, it is very difficult to view as constitutional—from our perspective—the political branches' prosecution of the Civil War. If a citizen/enemy against whom the government intends to direct military action in the context of lawful hostilities is entitled to due process before the government can deprive him of life or liberty, then the Union Army and Navy violated the Fifth Amendment every time they killed or captured a Confederate combatant. An argument that leads to the conclusion that Jefferson Davis was correct from our perspective in his argument with Abraham Lincoln over the constitutionality of secession and the Civil War is, I think, condemned by that conclusion alone. The constitutionality of the main elements of the war actions taken by the Lincoln administration and the Civil War Congresses is settled by political practice if anything is, in addition to being part of the judicial holding in the *Prize Cases* and elsewhere.[38] Interpreting *Hamdi* instead to introduce a significant but narrow rule requiring due process where the executive and a citizen disagree over the latter's enemy status for purposes of detaining the citizen respects the old precedents and maintains the proposition that it was Lincoln rather than Davis who got the Constitution right.[39]

38. This is not to say that there was an unmistakably clear right answer to the question of secession *before* the Civil War. The War is, I think, a good example of the sort of political branch decision that under Marshall's *McCulloch* reasoning on political practice is entitled to considerable weight in constitutional thought. On Davis's by no means irrational constitutional argument for secession's legality, *see* H. Jefferson Powell, *William Rawle and Secession: Legal Rights and Political Wrongs*, *in* Union and States' Rights: 150 Years After Sumter, A Legal History of Interposition, Nullification, and Secession (Neil H. Cogan ed., 2013).

39. But this may not be the correct reading of *Hamdi*, although I think it is the better one. Perhaps Justice O'Connor's opinion implies that a citizen the executive thinks is an enemy combatant but whose status is in doubt must always receive Fifth

It follows from all this that the geographic location of an enemy combatant against whom the executive decides to use military force is, in itself, constitutionally irrelevant. In early 2013, a U.S. Senator provoked a significant public controversy when he vigorously argued that the Constitution categorically prohibits any drone strike "against a U.S. citizen, on U.S. soil, and without trial."[40] Attorney General Eric Holder responded, first, that "[i]t is possible, I suppose, to imagine an extraordinary circumstance in which it would be necessary and appropriate under the Constitution . . . for the President to authorize the military to use lethal force within the territory of the United States," and subsequently, that the president has *no* " 'authority to use a weaponized drone to kill an American not engaged in combat on American soil.' "[41] The image evoked by the controversy is emotionally powerful, but the issue is a pseudo-problem. If the federal government is to have the power to wage war successfully, the war powers must permit the use of military force "on U.S. soil" and "against a U.S. citizen," if a citizen/enemy combatant who can

Amendment procedural due process before the executive can take his or her liberty for war purposes. It follows *a fortiori* that the executive must afford procedural due process in the same circumstances to someone whose life it wishes to take. Perhaps we could imply a limited exception for situations in which an urgent military action is necessary and there is no time to satisfy due process, but such an exception would be irrelevant for present purposes: The executive branch took many months to decide to kill al-Awlaki. If this second reading of *Hamdi* is correct, then in any situation where there is some reason to doubt that the citizen/target is in fact an enemy who may be and should be killed for war purposes, *Hamdi* requires procedural due process in advance of the drone strike, as the administration itself affirms. Unfortunately, as we shall see below, on that understanding of *Hamdi*, the administration's targeted killing policy fails to satisfy due process requirements and is unconstitutional.

40. Letter from Senator Rand Paul to John O. Brennan (Feb. 20, 2013), http://www.paul.senate.gov/files/documents/Brennan3.pdf.

41. Letters from Attorney General Eric H. Holder, Jr., to Senator Rand Paul (Mar. 4 and Mar. 7, 2013), http://www.paul.senate.gov/files/documents/BrennanHolderResponse.pdf, http://www.newyorker.com/news/amy-davidson/rand-paul-gets-a-letter-from-eric-holder.

lawfully be targeted is on American soil.[42] Once again, the Civil War illustrates and proves the point: The Confederate armies were made up, almost entirely, of U.S. citizens on U.S. soil. No doubt, and happily, such a situation is "extraordinary" in the literal sense of "not usual," but if it arises, there would be nothing constitutionally surprising about the conclusion that the executive could use lethal military force within the United States.[43]

C. FEDERAL WAR POWERS ARE DIVIDED

The Constitution distributes the plenary yet limited powers related to war between the two political branches.[44] This might seem obvious on the face of the constitutional text, but many people who write on the subject adopt (what is in essence if not expressly) the view that the war powers are not divided, that they are (in principle)

42. As future Chief Justice Hughes commented in his 1917 address, "[t]he power to use an army is co-extensive with the power to make war; and the army may be used wherever the war is carried on, here or elsewhere." Hughes, *supra* note 5, at 8.

43. The suggestion in Attorney General Holder's second letter that the citizen/enemy would have to be "engaged in combat" to be the object of a drone strike is inexact: Whether in the United States or abroad, an enemy combatant need not be literally engaged in combat at the time he is targeted in order to be a lawful target. International humanitarian law has no such requirement limiting the use of force against an enemy combatant. Admiral Yamamoto was not actually on a battlefield or directly engaged in immediate hostilities when killed.

44. The distribution of war powers is usually put this way, but the Article III branch also possesses powers, if subsidiary ones, related to war: Most obviously, the power to adjudicate charges of treason, the crime of "levying War against [the United States], or . . . adhering to their Enemies, giving them Aid and Comfort." Art. III, § 3, cl. 1. The federal courts' admiralty and maritime jurisdiction was a significant war-related power in the late eighteenth century. *See* William R. Casto, The Supreme Court in the Early Republic 71–125 (1995). Perhaps most important, judicial review of war decisions can play an important legitimating role by confirming the lawfulness of the government's actions. *See Boumediene*, 553 U.S. at 797 (judicial review "does not undermine the Executive's powers as Commander in Chief. On the contrary, the exercise of those powers is vindicated, not eroded, when confirmed by the Judicial Branch.").

entirely legislative (with the president reduced to administrator of the detailed military directions of Congress) or entirely executive (with Congress relegated to the tasks of paymaster and recruiting sergeant). It is my claim, instead, that the constitutional text and structure create a genuine division of substantive war powers between the two political branches, and that we need to get this background issue right if we are fruitfully to address the validity of the targeted killing of Anwar al-Awlaki or, indeed, anyone else.[45]

Article I, section 8 allocates to Congress a long list of powers, including most importantly for our purposes the power to declare war in clause 11, as well as the power to tax and spend "for the common Defence." Article II, section 2 begins with the statement that "[t]he President shall be the Commander in Chief of the Army and Navy of the United States." Congress's authority to exercise unenumerated or implied powers when doing so is "necessary and proper for carrying into Execution [its own enumerated] Powers" as well as "all other Powers vested by this Constitution in the Government of the United States," including (presumably) the president's commander in chief authority, which is obviously relevant, as is the president's implicit authority to make and pursue foreign policy.[46] On the face of it, the evident

45. In this book I sometimes refer to the Constitution's distribution of governmental authority among the branches of the federal governmental as the "separation of powers," but that familiar phrase is in fact misleading. As Professor Philip Bobbitt put it in an important essay, "the U.S. government ... is really a system of linked and sequenced powers. It makes far more sense to see the pattern of required cooperation in war as sequential, here as with other powers, in which one branch can act within certain boundaries, thereby having an impact on the choices open to the other branches but not determining the outcome of those choices." Philip Bobbitt, *War Powers: An Essay on John Hart Ely's War and Responsibility*, 92 MICH. L. REV. 1364, 1388 (1994).

46. As the Supreme Court observed in *Amer. Ins. Assoc. v. Garamendi*, "the source of the President's power to act in foreign affairs does not enjoy any textual detail, [but] the historical gloss on the 'executive Power' vested in Article II of the Constitution has recognized the President's 'vast share of responsibility for the conduct of our foreign relations.'" 539 U.S. 539 U.S. 396, 414 (quoting *Youngstown*, 343 U.S. 579, 610–11

purpose and effect of these features of the constitutional text is to break apart governmental authority over war matters, which in eighteenth-century English constitutional theory was vested almost entirely in the Crown (save for the power of the purse, a very important practical reservation); instead the American Constitution places a very extensive set of the disaggregated war powers in the legislature rather than the executive.[47]

I suggested in the previous paragraph that I think the most important clause in the provisions delegating military powers to Congress is the provision that "[t]he Congress shall have Power . . . To declare War." This claim needs justification because some constitutional scholars think that the authority conferred by the clause is limited to the power to issue formal declarations of war.[48]

(1952) (Frankfurter, J., concurring)). The president's chief tools in executing his or her foreign policy decisions are diplomatic, but we should not draw too sharp a line between authority over foreign affairs and authority with respect to military issues. Professor Matthew Waxman has persuasively argued that the president's power to threaten the use of military force is a deeply significant and constitutionally uncontested aspect of the president's authority over diplomacy, although Congress also possesses significant means for curbing the use of the power. *See* Matthew C. Waxman, *Syria, Threats of Force, and Constitutional War Powers*, 123 YALE L.J. ONLINE 297, 309–10 (2013) ("War powers decisions—in a practical sense, not a formal sense—occur earlier along the foreign policy decision tree than is generally acknowledged or understood in legal debates. Because the United States is a superpower that plays a major role in sustaining global security, its ability to threaten war is in some respects a much more policy-significant constitutional power than its power to actually make war.").

47. *See* Johnson v. Eisentrager, 339 U.S. 763, 788, 70 S. Ct. 936, 949 (1950) (citations omitted) ("Among powers granted to Congress by the Constitution is power to provide for the common defense, to declare war, to raise and support armies, to provide and maintain a navy, and to make rules for the government and regulation of the land and naval forces. It also gives power to make rules concerning captures on land and water, which this Court has construed as an independent substantive power. Indeed, out of seventeen specific paragraphs of congressional power, eight of them are devoted in whole or in part to specification of powers connected with warfare.").

48. *See, e.g.*, Robert J. Delahunty, *Structuralism and the War Powers: The Army, Navy and Militia Clauses*, 19 GA. ST. U. L. REV. 1021 (2003). *See especially id.* at 1055 n.127 (discussing the significance of the Declaration of War Clause if given this reading) .

Philip Bobbitt has argued that in the original eighteenth-century context, a declaration of war was not a condition precedent to military action: it was, instead, the "legal characterization of an ongoing conflict as one that is 'declared' or 'perfected,'" thereby "creating a juridical situation of total war." As such, the formal "declaration of war was a vestigial legal trait even at the time of the Constitutional Convention," and has played only a limited role in subsequent American constitutional history. The clause is not a nullity on this interpretation, he points out, because a declaration of war has effects "both in international law—as a declaration changes the status of nonbelligerents—and domestically—as it rallies the nation to a supreme effort," but the clause is in principle essentially irrelevant to questions about the initiation of hostilities or their conduct once begun.[49]

Despite the powerful textual and historical arguments that Professor Bobbitt has mustered, I do not think current constitutional thought should adopt his interpretation of the Declaration of War Clause. This is in part, as we shall see shortly, because the early Supreme Court appears to have adopted the broader construction of the clause followed in this book. But the more expansive reading of the Declaration of War Clause is, I think, the better interpretation in principle of a provision addressing a subject as important as war in a Constitution that is "the basic charter of our society" and "announces fundamental principles."[50] Put another way, interpreting the power "To declare War" as broad

49. Bobbitt, *War Powers*, supra note 45, at 1396-97. On Professor Bobbitt's interpretation, the founders expected, and history confirms, that the authorization of military action takes place by statutory authorization and through the interaction between Congress's powers to create the armed forces and appropriate funds, and the president's authority as commander in chief of the forces that Congress provides. *Id.* at 1390.

50. Poe v. Ullman, 367 U.S. at 540 (Harlan, J., dissenting); *Thornburgh*, 476 U.S. at 789 (White, J., dissenting).

authority to decide when the United States should initiate military conflict is more consonant with the Marshallian approach to constitutional thinking taken in this book. In a world where the founders already anticipated (correctly) that a formal declaration of war would seldom be necessary or even appropriate, *McCulloch v. Maryland* suggests that it would be an error to treat the power vested by the clause as narrow or formal in character. Consider the following.[51]

The Declaration of War Clause is located textually about midway through the list of congressional powers in Article I, section 8. Nothing about it suggests that it should be treated in any different fashion than that which we customarily apply to the other enumerated Article I powers—as a plenary grant of the power fully to execute Congress's policy preferences on the subject the clause addresses.[52] This suggests that Congress is empowered by the Declaration of War and Necessary and Proper Clauses to employ any means that are not constitutionally forbidden and that are conducive to the end of determining when, and whether, and with whom, the United States shall be at war.

51. *See* the very sensible discussion in *New Jersey Peace Action v. Obama*, 2009 WL 1416041 (D.N.J. 2009), at *8 ("The world of hostilities, therefore, does not neatly break down into the categories of "'war' and 'not war.' Rather, the political branches are empowered to conduct and maintain hostilities short of war and Congress may determine whether and when a declaration of war is necessary."), *aff'd*, 379 Fed. Appx. 217 (3d Cir. 2010), *cert. denied*, 131 S. Ct. 937 (2011).

52. Chief Justice Marshall's express references in *McCulloch* to the federal power over war are ambiguous on this score but only in the sense that he may have had in view the president's authority as commander in chief *in addition to* Congress's "great powers" over war. *See McCulloch*, 17 U.S. (4 Wheat.) at 407 ("among the enumerated powers of government ... we find the great powers, to lay and collect taxes; to borrow money; to regulate commerce; to declare and conduct a war; and to raise and support armies and navies. The sword and the purse, all the external relations, and no inconsiderable portion of the industry of the nation, are intrusted to its government."); *id.* at 411 ("the power of making war, or levying taxes, or of regulating commerce [is] a great substantive and independent power, which cannot be implied as incidental to other powers, or used as a means of executing them").

Even if we assume that the core meaning of the Declaration of War Clause is limited to the conferral of authority to decide whether to enter the state of total war created by an eighteenth century formal declaration, Congress is not limited to the core meaning of its enumerated powers in choosing what is the most "beneficial exercise" of those powers.[53] The power to decide on the use of military force in some other fashion than engaging in all-out war is clearly means conducive to the beneficial exercise of the power to declare such a war: doing so may obviate the need of the more drastic measure, or provide an alternative that in Congress's judgment is more likely to serve the Republic's needs. From the constitutional perspective outlined in *McCulloch v. Maryland*, it should be of no moment that Congress has chosen not to issue a formal declaration of war or has authorized the president to conduct only limited hostilities: the Constitution authorizes Congress to exercise its discretion on such matters. Congress's approval of lesser degrees of military hostilities is no less an exercise of the declare-war power than the regulation of some (but not all) interstate commerce is an exercise of the commerce power.

The fact that the power to declare war, like many others enumerated in Article I, section 8, intersects with and thus may affect the president's Article II power as commander in chief does not change the plenary nature of Congress's power. Chief Justice Rehnquist, writing for the Court in *Solorio v. United States*, made

53. *See McCulloch*, 17 U.S. (4 Wheat.) at 409. Later in his opinion, Chief Justice Marshall observed that "[i]t must have been the intention of those who gave these powers, to insure, so far as human prudence could insure, their beneficial execution. This could not be done, by confiding the choice of means to such narrow limits as not to leave it in the power of congress to adopt any which might be appropriate, and which were conducive to the end. This provision is made in a constitution, intended to endure for ages to come, and consequently, to be adapted to the various *crises* of human affairs." *Id.* at 415.

this point clearly with respect to another congressional power relating to military affairs.

> The constitutional grant of power to Congress to regulate the Armed Forces, Art. I, § 8, cl. 14, appears in the same section as do the provisions granting Congress authority, inter alia, to regulate commerce among the several States, to coin money, and to declare war. On its face there is no indication that the grant of power in Clause 14 was any less plenary than the grants of other authority to Congress in the same section.[54]

The reader should note that Rehnquist included the power to declare war in a list that can only have been intended to identify grants of plenary congressional authority.

Later in the *Solorio* opinion, Chief Justice Rehnquist rejected the argument that the Commander in Chief Clause somehow limits the scope or flexibility of Congress's power under the Military Government Clause. "[R]eading Clause 14 consistently with its plain language does not disserve [the constitutional goal of achieving the earlier English goal 'to transfer from the Crown to Parliament the control of the scope of court-martial jurisdiction']; Congress, and not the Executive, was given the authority to make rules for the regulation of the Armed Forces.... The unqualified language of Clause 14 suggests that whatever these concerns, they were met by vesting in Congress, rather than the Executive, authority to make rules for the government of the military." The language of clause 11 ("To declare War") is equally

54. 483 U.S. 435, 441 (1987). *Solorio* held that Congress has the constitutional authority to extend the jurisdiction of courts martial to members of the armed forces simply on the basis of their status, overruling an earlier decision limiting military jurisdiction to cases involving service-related charges.

unqualified, and there is no reason to treat its wording as any less of a plenary grant of power than the Military Government Clause or for that matter the Commerce Clause.[55] The argument that we should single out the Declaration of War Clause for narrower treatment than that we accord Congress's other enumerated powers has no basis in text, structure, or case law. The obvious constitutional purpose of transferring to Congress, not the president, much of the authority over war exercised by the Crown in Great Britain so that this Republic's commander in chief would possess a much more restricted area of discretion over military matters is ill-served by eviscerating the Declaration of War Clause of substantive force or by inflating the Commander in Chief Clause so that it essentially reconstitutes the pre-Revolutionary military prerogatives of the Crown.

This is not to say that it would be consistent with the Constitution's purposes to empty the Commander in Chief Clause of substantive force: The war powers are divided, not purely legislative.[56] From an early point in U.S. history, constitutional lawyers have pointed to the practical utility of locating the power to determine and order the execution of military strategy and tactics in a single officer rather in a two-house legislature. Partly as a matter of logic, and partly against the backdrop of the Revolutionary War, fought with the uniformed commander in chief subordinate to detailed instructions from a fractious and inexpert Congress, the Constitution has been understood to provide the Republic with the war-fighting advantages that a single,

55. *See also* Loving v. United States, 517 U.S. 748, 767 (1996) (the Military Government Clause "power is no less plenary than other Article I powers" and "[u]nder Clause 14 Congress, like Parliament, exercises a power of precedence over, not exclusion of, Executive authority.").

56. *See The President and the War Power: South Vietnam and the Sanctuaries*, 1 Op. O.L.C. Supp. 321, 324 (1970) (Rehnquist, Ass't Att'y Gen.) (the Commander in Chief Clause "is a substantive grant of power").

unified command structure makes possible.[57] The Commander in Chief Clause also serves a rather different purpose, the preservation of the nonmilitary character of the Republic: By ensuring that the commander in chief of the U.S. armed forces can only be the elected, civilian chief executive, the constitutional text puts a significant roadblock in the historically well-trodden path from generalissimo to military dictator.[58] As a matter of general constitutional reasoning—what Chief Justice Marshall called the "great paramount law of reason"—the best interpretation of the Constitution's text and purposes both recognizes Congress's primacy in deciding whether the United States will begin hostilities and if so what their intended scope should be *and* the president's robust authority to direct the conduct of lawful hostilities.

We need not rely on general reasoning alone for this view of the distribution of the federal war powers. Three very early, unanimous high Court decisions, two decided by opinions written by Chief Justice Marshall, are as a group impossible to explain except as showing that the Court of that era understood Congress to possess extensive powers over the initiation and scope of lawful military action. All three cases, *Bas v. Tingy, Talbot v. Seeman*, and *Little v. Barreme*, involved controversies that arose out of the use of force by a U.S. warship during the so-called Quasi-War with France

57. *See, e.g.*, 3 JOSEPH STORY, COMMENTARIES ON THE CONSTITUTION OF THE UNITED STATES 340–41 (1833).

58. *See, e.g.*, *Suspension of the Writ of Habeas Corpus*, 10 Op. Att'y Gen. 74, 79 (1861) (the president is commander in chief so that "whatever skilful soldier may lead our armies to victory against a foreign foe, or may quell a domestic insurrection; however high he may raise his professional renown, and whatever martial glory he may win, still he is subject to the orders of the civil *magistrate*, and he and his army are always 'subordinate to the civil power.'"). It is, to be sure, only a constitutional convention rather than a rule of the constitutional text that the president cannot be an active military officer. If the electoral process were to put into office a president in uniform, the Republic's problems would likely be ones well beyond the power of constitutional lawyers to address.

between 1798 and 1800. The Quasi-War began when American patience with high-handed French behavior and the depredations on American shipping of French privateers finally ran out. On July 7, 1798, Congress by statute abrogated the Revolutionary War treaty of alliance with France, and two days later Congress enacted a statute providing that the president was "hereby authorized to instruct the commanders of the public armed vessels which are, or which shall be employed in the service of the United States, to subdue, seize and take any armed French vessel, which shall be found within the jurisdictional limits of the United States, or elsewhere, on the high seas." President John Adams was also authorized to license privately owned American ships to act as privateers with "the same license and authority for the subduing, seizing and capturing any armed French vessel, and for the recapture of the vessels, goods and effects of the people of the United States, as the public armed vessels of the United States may by law have."[59] Subsequent legislation further expanded the hostile measures Congress authorized the executive to take against France, including the enforcement of a ban on commerce with France.

The following month, having "taken into consideration" both the congressional legislation and hostile French decrees of which he was aware, Attorney General Charles Lee provided Secretary of State Timothy Pickering a formal legal opinion, reasoning that "there exists not only an *actual* maritime war between France and the United States, but a maritime war *authorized* by both nations." Lee therefore concluded that "France is our enemy; and to aid, assist, and abet that nation in her maritime warfare will be treason in a citizen."[60] Lee's brief opinion leaves much to be inferred, but the significant constitutional assumptions in his

59. An Act Further to Protect the Commerce of the United States, ch. 68 & 69, 1 Stat. 578, 579 (July 9, 1798).

60. *Treason*, 1 Op. Att'y Gen. 84, 84 (1798).

thinking are clear: Congress may authorize hostilities by means other than a formal declaration of war, those hostilities are "war" for constitutional purposes (a citizen participating on the French side is a traitor and not simply an ordinary criminal), and the hostilities Congress authorizes may be less extensive than the complete set of hostilities on land and sea that a declaration would license (Congress has authorized a maritime war only).

The early Supreme Court cases, read together, adopt expressly and with some explanation and expansion Attorney General Lee's assumptions. The earliest of the three cases, *Bas v. Tingy*, involved the interpretation of congressional legislation.[61] On April 21, 1799, the USS *Ganges* under the command of Captain Thomas Tingey captured an American merchantman, the *Eliza*, that had been captured by a French privateer the previous month and was sailing with a French prize crew off the Canadian coast. The issue before the Court was the interpretation of a statutory provision that referred to vessels "re-taken from the enemy." The captain of the *Eliza* argued that because Congress had not declared war France was not "the enemy" within the meaning of the statute and thus Captain Tingey and the crew of the *Ganges* were not entitled to one-half the value of the *Eliza* and its cargo as "salvage" under the "re-taken" vessel provision. The Justices unanimously rejected that argument in seriatim opinions. Justice Samuel Chase's opinion provides the fullest statement of what seems clearly their common viewpoint.

> Congress is empowered to declare a general war, or congress may wage a limited war; limited in place, in objects, and in time. If a general war is declared, its extent and operations are only restricted and regulated by the jus belli, forming

61. 4 U.S. (4 Dallas) 37 (1800).

> a part of the law of nations; but if a partial war is waged, its extent and operation depend on our municipal laws. What, then, is the nature of the contest subsisting between American and France? In my judgment, it is a limited, partial, war. Congress has not declared war in general terms; but congress has authorised hostilities on the high seas by certain persons in certain cases. There is no authority given to commit hostilities on land; to capture unarmed French vessels, nor even to capture French armed vessels lying in a French port; and the authority is not given, indiscriminately, to every citizen of America, against every citizen of France; but only to citizens appointed by commissions, or exposed to immediate outrage and violence. So far it is, unquestionably, a partial war; but, nevertheless, it is a public war, on account of the public authority from which it emanates.[62]

Talbot v. Seeman, decided the following year with an opinion of the Court by the new Chief Justice, also involved an issue of salvage and squarely presented the question whether it was lawful for a Navy warship, the USS *Constitution*, to capture an originally German vessel previously seized by the French.[63] Chief Justice Marshall wrote that "[t]he whole powers of war being by the constitution of the United States, vested in congress, the acts of that body can alone be restored to as our guides in this enquiry. It is not denied, nor in the course of the argument has it been denied, that congress may authorize general hostilities, in which case the general laws of war apply to our situation; or partial hostilities, in which case the laws of war, so far as they actually apply to our

62. *Id.* at 43 (seriatim opinion of Chase, J.).
63. 5 U.S. (1 Cranch) 1 (1801).

situation, must be noticed."[64] *Talbot* thus confirmed, in an opinion representing the entire Court, Justice Chase's understanding of Congress's authority over the initiation of military action.

The last of these decisions, *Little v. Barreme*, involved the liability of Captain George Little of the USS *Boston* for the *Boston*'s allegedly unlawful capture of a ship, the *Flying-Fish*.[65] The merchantman was sporting the Danish flag, but Captain Little had reason to suspect that the *Flying-Fish* was in reality an American vessel attempting to circumvent the congressional prohibition on commercial intercourse with France. The legal difficulty arose from the fact that the *Boston* captured the *Flying-Fish* when the merchantman was sailing *from* a French port while the relevant statutory provision, read literally, authorized only the capture of vessels sailing *to* a French port. Captain Little's defense was that he was acting under the instructions of the secretary of the Navy, which directed him to capture suspicious vessels bound from French ports and that the secretary's instructions served what was clearly Congress's purpose better than would strict adherence to the letter of the statute. Chief Justice Marshall agreed with Little that the statute was poorly worded, and he suggested in dicta that in the absence of a statutory restriction the president might well have had the authority to order captures like that of the *Flying-Fish* in order to further the overall congressional purpose: To Marshall it was "by no means clear that the president . . . who is commander in chief of the armies and navies of the United States, might not, without any special authority for that purpose, in the then existing state of things, have [ordered the capture of] American vessels engaged in this illicit commerce." The suggestion, put differently, was that since Congress

64. *Id.* at 28.
65. 6 U.S. (2 Cranch) 170 (1804).

had authorized the president to carry out maritime hostilities against France, and had banned commerce with France as one aspect of those hostilities, he "might" have had the legal authority to order the Navy to capture vessels like the *Flying-Fish* as a means of pursuing Congress's purposes *as long as Congress had not restricted the commander in chief's discretion*.[66]

Unfortunately for Captain Little, Congress *had* restricted that discretion: The statute was, however unwisely, clear on its face, and the seizure, however sensible in terms of pursuing Congress's own war aims, was therefore unlawful: "[T]he legislature seem to have prescribed that the manner in which this law shall be carried into execution was to exclude a seizure of any vessel not bound to a French port." While it was true that Little was acting under "orders from the legitimate authority"—from "the department to which [the law's] execution was assigned," and that as a naval officer he was under a "general duty . . . to obey" such orders, his actions were nevertheless unlawful since executive orders "cannot change the nature of the transaction, or legalize an act which without those instructions would have been a plain trespass."[67]

66. 6 U.S. (2 Cranch) at 177 (citing the Take Care Clause of Article II, section 3). My sense from reading many Marshall opinions is that the convoluted syntax of his sentence about the president's commander in chief authority ("It is by no means clear that the president . . . might not . . .") masks a rather clear indication of Marshall's actual view, that the president *would* have had discretion to prosecute the congressionally authorized maritime war with France as he thought most effective in the absence of further congressional limitation.

67. *Id.* at 177–78, 179. Marshall commented that in light of "[t]hat implicit obedience which military men usually pay to the orders of their superiors," "the first bias of my mind was very strong in favor of the opinion that though the instructions of the executive could not give a right, they might yet excuse from damages." But, he explained, "I have been convinced that I was mistaken, and I have receded from this first opinion. I acquiesce in that of my brethren." *Id.* As a holding on the personal liability of government officials, *Little* has long since been overtaken by modern official immunity doctrine, which is much closer to Marshall's original view

Bas, *Talbot*, and *Little*, demonstrate that the early Supreme Court followed a consistent set of principles in construing the division of the war powers between Congress and the president.[68] Chief Justice Marshall's comment in *Talbot* that "the whole powers of war are by the constitution of the United States vested in congress" almost certainly referred, at least in part, to the power to determine whether to use military force and can serve as a shorthand summary for the unanimous view of the early Justices: Congress has plenary and exclusive authority to determine whether the United States shall initiate hostilities, and if so, what the limits and objectives of those hostilities are to be. Like most sweeping generalizations in law, this statement has a qualification: It would be unreasonable to worry that Marshall and his colleagues thought that the president must await an act of Congress before employing military force in response to an attack on the United States. In terms of the history laid out in Chapter One, this is the constitutional circumstance that applied in Phase Two of the current conflict, and as I argued there, from the beginning constitutional lawyers have assumed that the executive may

than to that of his brethren. What is significant for our purposes is that Marshall apparently had was certain that the president had no power to authorize Captain Little to use military force beyond the limitations that Congress had imposed in authorizing the executive to act. This is unsurprising in light of Marshall's earlier opinion in *Talbot v. Seeman*, but it goes a very long way to confirming that Marshall, and his brethren, decided these cases on the basis of a thought-out and clearly held understanding of the relative powers of Congress and the president with regard to the conduct of hostilities.

By "the department to which its execution was assigned" Marshall is likely to have meant the executive branch rather than the Department of the Navy specifically, but that is immaterial as a constitutional matter: Throughout his opinion Marshall assumes that the secretary's orders are to be treated as if they were the president's in his analysis.

68. *See also* Brown v. United States, 12 U.S. (8 Cranch) 110, 128–29 (1814) (Marshall, C.J.) (decision whether to confiscate enemy property as a war measure is a policy matter for Congress exercising "the sovereign power" to express "the will of the nation").

respond to attack. But we are no longer in Phase Two. Congress has acted, and the relevant constitutional principle according to the early Supreme Court decisions is that Congress has the presumptive authority to set whatever limitations it thinks appropriate on the use of military force—even when, as was the case with the Quasi-War, the other country has initiated hostilities and the United States is acting in self-defense. In Justice Chase's words, "Congress is empowered to declare a general war, or congress may wage a limited war; limited in place, in objects, and in time," or as Chief Justice Marshall put it, "congress may authorize general hostilities . . . or partial hostilities" as it thinks most appropriate.

The early cases do not address the questions whether a "general war" must as a constitutional matter be authorized by a formal declaration of war, or the extent to which the authorization of a "general war" would confer on the president as commander in chief greater discretion in the conduct of the war than he otherwise would enjoy. It is clear, I believe, that the early Justices did not distinguish "war" and "hostilities" for the purposes of constitutional analysis: Congress's "whole powers of *war*" encompassed for them the power to authorize the use of military force at any level and, at least short of "general war," to limit the commander in chief's discretion as to the "place, objects, and time" within which he is authorized to act. Whatever President Adams's legal options might have been in the absence of Congress's limited authorization for the use of force (in the parallel situation to what I have termed Phases One and Two of the current conflict), once Congress authorized certain hostilities and by implication forbade others, Adams was legally constrained to observe Congress's limitations. Captain Little's actions were *contra legem*, contrary to the governing statutory law as the Court construed that law, and could not be rendered lawful by a presidential decision based on Article II alone, no matter how sensible Adams's decision might be, even viewed as a means of pursuing the policy goals of *Congress* in authorizing the Quasi-War.

As noted earlier, U.S. constitutional law has no doctrine of decisional desuetude, and none of these three early cases have been overruled or even put in question by later Supreme Court decisions as to their reasoning about the distribution of war powers. On the other hand, the Court decided the cases long ago and has seldom referred to them since then, there was no opinion of the Court in *Bas* (though there was no apparent disagreement on the constitutional issues either), and Chief Justice Marshall's opinion in *Talbot* handled the constitutional issue in a clear but cursory fashion. As a formal matter the cases' authority seems unimpaired, but standing on their own a scholar or legal advisor might hesitate to put too much weight on them simply as authority, however much they might bind a federal district judge. Other sources of constitutional argument follow the basic constitutional reasoning in *Bas, Talbot*, and *Little*, however, and support the conclusion that we should treat them as authoritative. Perhaps the most significant are two opinions written in 1970 by then–Assistant Attorney General William H. Rehnquist, both of which defend the lawfulness of President Nixon's decision in the spring of that year to deploy U.S. ground and air forces into Cambodia for the purpose, according to the administration, of destroying the base camps that North Vietnamese and Viet Cong forces were using as supply depots and sanctuaries in that officially neutral country.[69]

69. *See The President and the War Power: South Vietnam and the Cambodian Sanctuaries*, 1 Op. O.L.C. Supp. 321 (1970) [*Cambodian Sanctuaries*]; *Presidential Authority to Permit Incursion into Communist Sanctuaries in the Cambodia-Vietnam Border Area*, 1 Op. O.L.C. Supp. 313 (1970) [*Authority to Permit Incursion*]. Assistant Attorney General Rehnquist's opinions provide, to the best of my knowledge, the most extensive analysis of the constitutional distribution of war powers in *any* executive branch legal opinion in history. They are significant for the level of care and detail they display, the objectivity and fair-mindedness of the discussion, and

Assistant Attorney General Rehnquist's conclusion that the Cambodian sanctuaries operation was lawful rested on a narrow and carefully nuanced line of reasoning. The controlling premise of Rehnquist's argument was that Congress had authorized the war. U.S. armed forces were fighting in South Vietnam and in the skies over North Vietnam. No act of Congress purported to prohibit the conduct of operations in Cambodia as part of those congressionally authorized hostilities; Rehnquist noted that in contrast Congress had enacted prohibitions on the conduct of hostilities in Laos and Thailand and that the executive branch accepted those prohibitions.[70] As a factual matter, the judgment of the president and his military advisors that the enemy's ability to use sanctuaries located in Cambodia but within a few miles of South Vietnam (and indeed not far from Saigon) without fear of U.S. or South Vietnamese attack provided the enemy with a distinct tactical advantage, putting U.S. (and South Vietnamese) personnel at greater risk than necessary and interfering significantly with the successful achievement of the objectives identified

of course, the identity of their author. I discuss the Cambodian sanctuaries opinions in detail in H. JEFFERSON POWELL, THE PRESIDENT AS COMMANDER IN CHIEF: AN ESSAY IN CONSTITUTIONAL VISION 136–47 (2014), and I will not repeat that entire analysis here.

70. *Cambodian Sanctuaries*, 1 Op. O.L.C. Supp. at 334. The reasoning of Rehnquist's opinions would allow, I think, for the conclusion that a geography-defined prohibition on the use of military force might be unconstitutional as applied if the president reasonably concluded that disregarding the prohibition temporarily was necessary in order to protect American lives or pursue the legislatively defined objectives of the congressionally authorized hostilities. On the other hand, *Little v. Barreme*, the authority of which Rehnquist fully accepted, would make it impossible to justify a presidential decision to set aside a geographic restriction across the board on the ground that the president thought Congress's restriction was unwise or self-defeating. The Adams administration thought the same about the statutory limitation at issue in *Little*, and Marshall, who agreed with the administration on the imprudence of the limitation, squarely rejected the idea that the executive's judgment could provide a lawful basis for the general order under which Captain Little acted.

by Congress in the Gulf of Tonkin resolution authorizing the war. The operation did not involve the conduct of hostilities against Cambodian armed forces, and the Cambodian government's protest was purely pro forma: The (new) Cambodian government was acting as a de facto ally and in any event was incapable itself of fulfilling its international law obligation to prevent enemy use of its territory. Finally, the operation's tactical purposes and anticipated duration made it virtually impossible that it would lead to an expansion of the existing, authorized war beyond the limited period in which it would be taking place.[71]

The fact that an executive branch legal opinion approves the legality of a presidential decision is sometimes treated as, in effect, prima facie evidence that the opinion is badly reasoned or insincere.[72] That one may rightly exercise caution when someone's lawyers provide an opinion of counsel that upholds the lawfulness of their client's actions is mere common sense, but this caution is misused when it serves as an excuse not to examine the president's lawyers' reasoning for its intrinsic persuasiveness: The best argument one can make for a client, after all, is one that others will conclude is correct. As *McCulloch v. Maryland* illustrates, the supreme law of the Constitution is vindicated as fully when political branch action is sustained as when the

71. There is a valuable discussion of the factual predicates for Rehnquist's opinions in Timothy Guiden, *Defending America's Cambodian Incursion*, 11 ARIZ. J. INT'L & COMP. L. 215, 228–30 (1994). I discuss the possibility that there were covert purposes behind the Cambodian operation that Nixon and "all the President's men" did not disclose to the Justice Department in POWELL, *supra* note 69, at 145 & nn.24 & 25. While I do not credit the logical possibility that Rehnquist was himself privy to these darker designs (if they in fact existed), even if he had been, the significance of his opinion would lie in his reasoning, not in his personal motives.

72. Even some of the more thoughtful critiques of the executive branch legal opinions on the killing of al-Awlaki appear to fall into this logical error. *See, e.g.*, Glenn Greenwald, Chilling Legal Memo from Obama DOJ Justifies Assassination of US Citizens (Feb. 5, 2013), http://www.theguardian.com/commentisfree/2013/feb/05/obama-kill-list-doj-memo (discussing the DOJ White Paper).

constitutional lawyer must conclude that the action is unconstitutional. My judgment is that both the specific reasoning and the specific conclusion of the Cambodia opinions are persuasive, but their real importance lies elsewhere, in Rehnquist's decision to embed his resolution of the specific question he was asked in a discussion of the constitutional distribution of war powers that explicitly and expressly rejected any sort of presidential-power triumphalism. Rehnquist instead endorsed a view of the war powers entirely consistent with that adopted long before by the Supreme Court in the three early decisions I have discussed.[73]

The constitutional text, Rehnquist insisted, is structured to provide for "[t]he accommodation of . . . two interests—the prohibition of one-man commitment of a nation to war and the need for prompt executive response to international situations. . . . The Founding Fathers intended to distinguish between the initiation of armed conflict, which is for Congress to determine, and armed response to conflict situations, which the Executive may undertake."[74] The result of this deliberate division of power and responsibility, Rehnquist reasoned, was a history of practical accommodations: "Nearly two hundred years of practice under the constitutional system ha[ve] given rise to a number of precedents and usages."[75] Rehnquist resisted the temptation to read American political practice as unequivocally supportive

73. This was not mere coincidence: The Cambodian sanctuaries opinions expressly treated *Bas* and *Little* as constitutional authorities. *See Cambodian Sanctuaries*, 1 Op. O.L.C. Supp. at 323–25; *Authority to Permit Incursion*, 1 Op. O.L.C. Supp. at 314–16.

74. *Authority to Permit Incursion*, 1 Op. O.L.C. Supp. at 314. Rehnquist's reasoning in both opinions clearly rejects a minimal interpretation of the Declaration of War Clause that would reduce the authority it delegates to the power to invoke an international law status. *See Cambodian Sanctuaries*, 1 Op. O.L.C. at 332 ("constitutional practice must include executive resort to Congress in order to obtain its sanction for hostilities which reach a certain scale").

75. *Id.* at 322.

of unilateral presidential authority. Unlike partisans of either branch's power, Rehnquist candidly admitted that "[t]he precise number of [unilateral presidential actions] is a matter of some dispute, as is the legitimacy of some of them," and that "[t]here has been dispute, often bitter, as to how far the President may go alone on his constitutional authority."[76] In particular, he was evidently unsure how to treat Korea: He was unwilling simply to dismiss it as aberrational or a massive constitutional error on President Truman's part, but he was uneasily aware that it could scarcely be viewed as the exercise of a limited Article II power to make short-term, low-intensity use of military force.[77]

Despite its complications, Rehnquist argued that the history of political practice makes it appropriate to recognize the existence of some degree of presidential power to employ military force without statutory authorization, not only in the uncontroversial circumstance of an attack on the United States (a concept itself fuzzy on the edges, as we have seen) but also as a tool of foreign policy. The opposite conclusion, that only the emergency created by an attack justifies nonstatutory presidential action, is a poor fit with U.S. constitutional history and disregards the claim of right under which presidents have made unilateral use of force, the reasoned discussions in Congress and between the political branches at various times about the validity of that claim, and Congress's general acquiescence over time. Practice has accommodated the two constitutional "interests" (I might say "purposes") that Rehnquist identified by allowing for

76. *Authority to Permit Incursion*, 1 Op. O.L.C. at 315, 317–18.

77. *See id.* at 316–17 (discussing "the Great Debate" in Congress over the lawfulness of U.S. involvement in the Korean War, and concluding abruptly that "it is clear that Congress acquiesced in the President's action"); *Cambodian Sanctuaries*, 1 Op. O.L.C. at 329 (admitting that Truman's invocation of the UN Charter could not "obviate a . . . requirement" of congressional authorization but concluding that "Congress acquiesced").

relatively brief, low-intensity military actions on presidential initiative while requiring congressional approval for hostilities that are (likely to be) lengthy, involve large-scale military operations, or entangle the United States in wider geopolitical conflicts.[78] The Vietnam War itself, he essentially conceded, clearly required congressional authorization in the light of its "duration . . . and its requirements in terms of both men and material."[79]

Assistant Attorney General Rehnquist's Cambodia opinions are a clear example of a senior executive branch lawyer—and a most distinguished one—adhering to a nuanced and historically grounded account of the relationship between the congressional and presidential roles in the initiation of armed conflict by the United States that avoids untenable claims of presidential supremacy. Rehnquist's views have been followed by some, though by no means all, subsequent Office of Legal Counsel (OLC) opinions in drawing a constitutional distinction between the limited hostilities a president may undertake on the basis of Article II alone and the more intense sorts of military action

78. *See Authority to Permit Incursion*, 1 Op. O.L.C. at 317, 314–15 (Congress has "the sole authority" to decide on "all-out war," and quoting *Bas v. Tingy* for the distinction between that level of hostilities and "limited conflict"). Rehnquist recognized that the president can effectively force Congress's hand by taking actions that embroil the United States in conflict before Congress can make a judgment, but he insisted that the president has a good faith duty not to circumvent Congress's role.

> It is too plain from the foregoing discussion to admit of denial that the Executive, under his power as Commander-in-Chief, is authorized to commit American forces in such a way as to seriously risk hostilities, and also to actually commit them to such hostilities, without prior Congressional approval. However, if the contours of the divided war power contemplated by the framers of the Constitution are to remain, constitutional practice must include Executive resort to Congress in order to obtain its sanction for the conduct of hostilities which reach a certain scale. Constitutional practice also indicates, however, that Congressional sanction need not be in the form of a declaration of war.

Cambodian Sanctuaries, 1 Op. O.L.C. Supp. at 333.
79. *Id.* at 335.

("'war' in the constitutional sense") that require congressional authorization if initiated by the United States.[80]

Rehnquist's Cambodia opinions thus rejected what we might call Article II supremacism as to the power to initiate hostilities. He just as clearly rejected the other strand of supremacist thought: the claim that the president has complete and exclusive discretion over specific military decisions in the conduct of lawful hostilities. Rehnquist quoted a familiar passage from Chief Justice Chase's concurrence in Ex parte *Milligan* that asserts that Congress's powers "not only to raise and support and govern armies, but to declare war'" provide it the constitutional authority "'to provide by law for carrying on war *except such as interferes with the command of the forces and conduct of campaigns [which] belong to the President as commander in chief.*'"[81] Unlike the early Supreme Court decisions, which were silent on the question, Rehnquist asserted the existence of a core of exclusive presidential commander in chief authority, a view in line with the longstanding views of the presidents and attorneys general. But as with the power to initiate hostilities, Rehnquist proposed a constitutional position that sought to reconcile the constitutional

80. *See, e.g., Deployment of United States Armed Forces into Haiti*, 18 Op. O.L.C. 173, 178 (1994). "In our view, determining whether a particular planned engagement constitutes a 'war' for constitutional purposes instead requires a fact-specific assessment of the 'anticipated nature, scope, and duration' of the planned military operations." *Authority to Use Military Force in Libya*, 2011 WL 1459998, at *8 (O.L.C. Apr. 1, 2011). Between the attacks of 9/11 and early 2003, several OLC opinions adopted a view of the president's commander in chief power that treated it as essentially unbounded by Congress's powers, but the office essentially disavowed those opinions at the end of the Bush administration. *See Status of Certain OLC Opinions Issued in the Aftermath of the Terrorist Attacks of Sept. 11, 2001*, 2009 WL 1267352, at *4 (O.L.C. Jan. 15, 2009) ("Article I, Section 8 of the Constitution also grants significant war powers to Congress").

81. *Cambodian Sanctuaries*, 1 Op. O.L.C. at 331 (quoting Ex parte *Milligan*, 71 U.S. (4 Wall.) 2, 139 (Chase, C.J., concurring in the judgment) (1866)) (emphasis supplied by Rehnquist).

principle articulated in the early cases with a history of political practice that seemed to legitimize some degree of autonomous commander in chief authority.

Rehnquist invoked *Little* as support for the proposition that "Congress undoubtedly has the power in certain situations to restrict the President's power as Commander in Chief to a narrower scope than it would have had in the absence of legislation." Rehnquist's specific comments about Congress's power to limit the discretion of the president as commander in chief substantially tracked the reasoning in *Bas* and *Little*: Congress need not authorize "all-out war" and in doing so give the president "in effect a blank check . . . to conduct military operations to bring about the subjugation" of the enemy. It may instead "delegate a lesser amount of authority" to the president—the early Supreme Court's concept of "limited war" (*Bas*) or "partial hostilities" (*Talbot*). By authorizing only limited hostilities, Congress can "retain[] a larger degree of control" over the president's military choices. Rehnquist echoed Justice Chase in *Bas* in specifically acknowledging Congress's constitutional authority to impose geographical limitations and to define "the objectives for which the conflict may be carried on." Rehnquist gave the Vietnam War itself, despite its scale, as an example of congressionally limited hostilities. Congress had exercised its power to regulate the commander in chief's discretion in its initial authorization, which specified the objectives of the hostilities Congress was allowing the president to pursue, and by subsequently forbidding the deployment of U.S. armed forces to Laos or Thailand "in connection with the Vietnam conflict."[82]

Assistant Attorney General Rehnquist closely followed the early Supreme Court cases on Congress's power over war matters

82. *Id.* at 334.

while affirming one position that the early Court did not address, the possibility of unconstitutional interference with the president's authority "should Congress attempt by detailed instructions as to the use of American forces already in the field to supersede the President as Commander in Chief."[83] Rehnquist's approach to the president's constitutional role respects the language and evident purpose of the Commander in Chief Clause *and* beyond that reservation gives full effect to the usual principle that the powers delegated to Congress afford the legislative branch plenary regulatory authority within their respective domains. On the level of practice, while this view does not render all questions easy—no general view can—it respects Justice Jackson's admonition that "[w]hile the Constitution diffuses power the better to secure liberty, it also contemplates that practice will integrate the dispersed powers into a workable government."[84] Whatever answer we give to the constitutional problem of targeted killing, it must satisfy both constitutional principle and practical workability if it is to be satisfactory.

83. *Id.* at 334.
84. *Youngstown*, 347 U.S. at 635 (concurring opinion).

Chapter 4

The Killing of Anwar al-Awlaki: A Constitutional Analysis

We are now ready to think constitutionally about the targeted killing of Anwar al-Awlaki. The decision to kill him was, necessarily under the Constitution of the United States, a matter for the executive branch, and the first constitutional question to be asked was whether the president had the affirmative legal power to make the decision or allow his subordinates to do so. The killing took place during what Chapter One termed Phase Three of the war on terror, and so the scope of the president's authority to take military action against al-Qaeda is affected by Congress's passage of the 2001 Authorization for the Use of Military Force (AUMF). Only if the decision was constitutionally authorized is it necessary to ask the second question, whether killing al-Awlaki violated an applicable constitutional prohibition.

A. A PRELIMINARY ISSUE: OUR SOURCES FOR THE ADMINISTRATION'S CONSTITUTIONAL THINKING

Unlike the views of the Clinton and Bush administrations on the president's authority to act against terrorists during Phases One and Two, which are not public, we do not have to speculate about the Obama administration's constitutional thinking about the killing of al-Awlaki. A few days after the drone strike, a reporter for the *New York Times*, Charlie Savage, reported that anonymous sources had disclosed to him the existence of "a roughly 50-page memorandum by the Justice Department's Office of Legal Counsel (OLC), completed around June 2010." The OLC Memo, as Savage put it, "provided the justification for acting [against al-Awlaki] despite an executive order banning assassinations, a federal law against murder, protections in the Bill of Rights and various strictures of the international laws of war."[1] As reported subsequently, the 2010 memo expanded on earlier OLC advice that the killing of al-Awlaki would be lawful.[2] As we now know, the formal date of OLC's second and more elaborate legal opinion on targeting al-Awlaki was July 16, 2010, fourteen months before the drone strike that killed him.

For over three years, however, the Obama administration refused to release the OLC legal opinion, or indeed even to

1. *See* http://www.nytimes.com/2011/10/09/world/middleeast/secret-us-memo-made-legal-case-to-kill-a-citizen.html?pagewanted=all&_r=0 (Oct. 8, 2011) The Office of Legal Counsel, "OLC" in government-speak, is the component of the U.S. Department of Justice that carries out the attorney general's constitutional and statutory duties of providing legal advice to the president and other high executive officials. Savage, who had not seen the OLC opinion himself, was informed that "[t]he memo . . . was narrowly drawn to the specifics of Mr. Awlaki's case and did not establish a broad new legal doctrine to permit the targeted killing of any Americans believed to pose a terrorist threat." In fact, it became clear from subsequent speeches by the president, Attorney General Eric Holder, and other high executive branch officials, that the Obama administration was following the legal analysis described to Savage in carrying out its overall policy of targeted killing as a tool in the conflict with al-Qaeda and its allies. *See generally* Kenneth Anderson & Benjamin Wittes, *Speaking the Law: The Obama Administration's Addresses on National Security Law* (2013-on) (providing a comprehensive survey of the administration speeches and showing their internal consistency).

2. According to reports published in the *New York Times* and elsewhere, OLC drafted a short memo concluding that it would be lawful to kill al-Awlaki in the aftermath of "the attempted bombing of a Detroit-bound airliner on Dec. 25, 2009," and wrote the lengthier July 2010 opinion in response to concerns raised by public discussions of 18 U.S.C. § 1119(b) (making the unlawful killing of a U.S. citizen

confirm its existence, although subsequent events vindicated Savage's report. But in a remarkable series of public addresses, administration officials defended the legality of the administration's employment of targeted killing along lines consistent with Savage's initial story. The highest-profile official discussion of the administration's views was delivered by President Obama himself on May 23, 2013, in a major speech on national security policy. He looked forward to an end, eventually, to the state of authorized (if undeclared) war that has existed since September 11, 2001, and specifically to an end to the policy of killing specific, targeted individuals that the government has implemented in his and the preceding administrations. The president vigorously asserted the administration's commitment to respecting the Constitution and justified the legitimacy and specifically the legality of the killing of al-Awlaki and the other individuals whom the United States has targeted.[3] The president's defense included the sweeping assertion that "[f]or the record, I do not believe it would be constitutional for the government to target and kill any U.S. citizen—with a drone, or a shotgun—without due process." The administration's internal processes for deciding to carry out a targeted killing, he insisted, address this constitutional requirement. He went on at once to assert that "when a U.S. citizen goes abroad to wage war against America—and is actively plotting to kill U.S. citizens; and when neither the United States, nor our partners are in a position to capture him before he carries out a plot—his citizenship should no more serve as a

abroad a federal crime). *See* http://www.nytimes.com/2013/03/10/world/middleeast/anwar-al-awlaki-a-us-citizen-in-americas-cross-hairs.html; http://www.nytimes.com/2014/06/24/us/justice-department-found-it-lawful-to-target-anwar-al-awlaki.html .

3. *See* President Barak Obama, The Future of Our Fight Against Terrorism (National Defense University, May 23, 2013), http://www.aclu.org/files/assets/obama-speech.pdf .

shield than a sniper shooting down on an innocent crowd should be protected from a swat team." The killing of al-Awlaki, who "was continuously trying to kill people," was therefore legal and constitutional.[4]

The reasons for the Obama administration's reluctance to disclose the legal analysis in the 2010 OLC Memo are unclear. The administration's concern seems to have gone beyond the reflexive executive branch preference for holding close information which might shine a light on the executive's internal deliberations. And yet nothing in the 2010 document is inconsistent with the public case for the administration's targeted killing policy that the president and others laid out months before. Nor could there be any sound reason for the administration to resist confirming that its lawyers had addressed questions about the legality of killing al-Awlaki and others: The fact that President Obama and his legal advisors concerned themselves about whether he had the power to employ drones as a means of killing targeted individuals, including those who are U.S. citizens, and specifically the power to kill al-Awlaki, is all to the good from any perspective. We need not speculate, however, about the reasons for the Obama administration's reluctance to disclose the general contents of its lawyers' constitutional thinking, and in any event, the advice eventually came out.[5] In early 2013, an unsigned Department of Justice "white paper" that summarized the 2010 memorandum

4. RLINK, http://www.aclu.org/files/assets/obama-speech.pdf . The president explained that he "would have detained and prosecuted Awlaki if we captured him before he carried out a plot. But we couldn't. And as President, I would have been derelict in my duty had I not authorized the strike that took out Awlaki." *Id.*

5. It is easy to understand why the administration might have preferred to keep any details about its knowledge of al-Awlaki's activities confidential: Such details might inadvertently reveal to al-Qaeda or others currently secret U.S. modes of intelligence gathering. What is puzzling is its reluctance to release a version of the 2010 memorandum redacted so as to protect those aspects of the document.

was leaked to NBC reporter Michael Isikoff.[6] Finally, on June 23, 2014, the U.S. Court of Appeals for the Second Circuit released a redacted version of the 2010 memo after the administration decided not to appeal an adverse decision in a Freedom of Information Act case seeking the memo's disclosure.[7]

B. THE CONSTITUTIONAL AUTHORIZATION FOR THE KILLING OF ANWAR AL-AWLAKI

1. The Administration's Position

Neither the redacted OLC Memo nor the leaked DOJ White Paper addresses the question of authorization at great length. The White Paper summarizes the position of the administration's lawyers usefully:

> The United States is in an armed conflict with al-Qa'ida and its associated forces, and Congress has authorized the President to use all necessary and appropriate force against those entities. *See* Authorization for Use of Military Force ("AUMF"), Pub. L. 107-40 § 2(a), 115 Stat. 224, 224 (2001). In addition to the authority arising from the AUMF, the

6. *See* Department of Justice White Paper, *Lawfulness of a Lethal Operation Directed Against a U.S. Citizen Who Is a Senior Operational of Al-Qa'ida or an Associated Force* (unsigned and undated) [henceforth, DOJ White Paper]. The White Paper was not a formal legal opinion; the journalist who revealed it reported that "administration officials" described it "as a policy document," http://openchannel.nbcnews.com/_news/2013/02/04/16843014-justice-department-memo-reveals-legal-case-for-drone-strikes-on-americans?lite.

7. *See Applicability of Federal Criminal Laws and the Constitution to Contemplated Legal Operations Against Shaykh Anwar al-Aulaqi* (July 16, 2010), https://www.justsecurity.org/wp-content/uploads/2014/06/OLC-Awlaki-Memo.pdf [henceforth, OLC Memo]. There is no discernible difference in constitutional analysis between the two Justice Department documents, and so I shall draw on them both. The White Paper is in fact often more coherent because it has not been redacted.

> President's use of force against al-Qa-ida and associated forces is lawful under other principles of U.S. and international law, including the President's constitutional responsibility to protect the nation and the inherent right to national self-defense recognized in international law. It was on these bases that the United States responded to the attacks of September 11, 2001, and "[t]hese domestic and international legal authorities continue to this day."[8]

The White Paper makes three legal claims in this paragraph. First, it refers to four distinct legal sources for the president's authority to use force against al-Qaeda and its allies in response to the events of 9/11: (a) the AUMF (statutory authorization), (b) unspecified "other principles of U.S. and international law" in addition to (c) "the President's constitutional responsibility to protect the nation," and (d) "the inherent right to national self-defense recognized in international law." Second, the White Paper implies that there are other sources of legal authority. "In addition to the authority arising from the AUMF, the President's use of force . . . is lawful under *other principles of U.S. and international law, including*" the president's "constitutional responsibility" and the national right of self-defense. Third, the White Paper asserts that all of these legal sources of authority "'continue to this day,'" which appears to imply that the enactment of the 2001 AUMF—in terms of Chapter One, the transition from Phase Two to Phase Three—had no legal effect on the scope of the president's authority, at least with respect to actions such as the use of targeted killing against al-Qaeda. I discuss the four asserted sources of authorization for the targeted killing policy in a different order than the White Paper presents them, because

8. DOJ White Paper, *supra* note 6, at 2 (omitting some citations).

the correct analysis of a claim to rest the policy on the president's "constitutional responsibility" depends on understanding what Congress has done in enacting the AUMF.

2. International Law as Authorization for the President's Policy

There is no doubt that under international law "the United States" was entitled to respond militarily to al-Qaeda's aggression in some fashion. But the president is not "the United States," and international law does not purport to specify which branch of the U.S. government is entitled as a matter of domestic constitutional law to exercise the Republic's international law right of national self-defense. As a matter of U.S. domestic law, furthermore, we can presume that at least one of the branches of the federal government is empowered to exercise any sovereign power recognized as legitimate by international law including the right of self-defense,[9] although the presumption should not be pressed too far.[10] But such a presumption says nothing about whether and when and to what extent *the president* may exercise the international law right on behalf of the United States. The constitutional text demonstrates that at least some of the domestic legal authorities relevant to exercising the international law right are delegated to Congress rather than the executive: Article I empowers Congress, not the president, to declare war, issue

9. *See, e.g.*, Fong Yue Ting v. United States, 149 U.S. 698, 711–12 (1893) (U.S. government possesses all "inherent and inalienable right[s] of every sovereign and independent nation" while noting the division of the relevant powers between the president and Congress).

10. As we have seen, the Supreme Court has repudiated the notion that the U.S. government has powers that derive from any source other than the Constitution, directly or indirectly. The cases looking to international law in defining the powers incident to national sovereignty that the U.S. government may exercise (as in *Fong Yue Ting*) are best read as looking to international law in interpreting the

letters of marque and reprisal, establish the armed forces, provide for federalizing the National Guard to "repel Invasions," and (by inference) suspend habeas corpus "in Cases of . . . Invasion [when] the public Safety may require it";[11] the third amendment assumes that it is Congress that has the power in wartime to co-opt private housing to quarter troops.[12] In discussing whether *the president* has the domestic legal authority to carry out the targeted killing policy, almost nothing is gained by observing that *the United States* has that authority under international law.

International law cannot of its own force authorize the president to do anything at all if we are asking about his or her authority to act under U.S. domestic law. The specific question whether international law can provide the president with domestic law

constitutional powers of the government—in other words, there is a presumption that the Constitution has vested internationally recognized national powers in the U.S. government. But the presumption does not support any argument that international law itself provides domestic law authorization or allows the government to take actions contrary to constitutional prohibitions such as the Bill of Rights. *See, e.g.*, Amer. Ins. Assoc. v. Garamendi, 539 U.S. 396, 416 n.9 (2003) (citation omitted) ("Even Justice Sutherland's reading of the National Government's 'inherent' foreign affairs power in *United States v. Curtiss-Wright Export Corp.*... contained the caveat that the power, 'like every other governmental power, must be exercised in subordination to the applicable provisions of the Constitution.'").

11. U.S. CONST. art. I, §§ 8 & 9. The passive voice construction in the Suspension Clause of Article I, section 9 gave President Lincoln and Attorney General Bates an opening to argue that the president has the authority to suspend the writ unilaterally in at least some circumstances. Chief Justice Taney expressly rejected any such claim in Ex parte *Merryman*, 17 F. Cas. 144 (C.C.D. Md. 1861) noting correctly that section 9 appears on its face and by its placement in the Constitution to be a set of restrictions on Congress's powers. *Boumediene v. Bush*, 553 U.S. 723 (2008), did not discuss Lincoln's argument and strongly implied that Taney was right and the power of suspension belongs to Congress.

12. The Third Amendment permits the quartering of troops in private houses "in time of war, but in a manner to be prescribed by law," in other words, as *Congress* prescribes. As Justice Jackson pointed out in *Youngstown*, "[t]ime out of mind, and even now in many parts of the world, a military commander can seize private housing to shelter his troops. Not so, however, in the United States." 343 U.S. at 644 (Jackson, J., concurring).

authority to take military action was squarely presented at the start of the Korean War. In 1950, President Truman unilaterally committed American troops to the defense of South Korea in response to the North Korean invasion. Truman invoked the call by the U.N. Security Council for an immediate ceasefire in his June 27 announcement of the U.S. intervention, and on the same day the Security Council adopted a further resolution recommending to member states that they provide assistance in repelling the invasion. Truman and his administration nonetheless quickly retreated from the suggestion that action by the Security Council could provide domestic law authorization for the president's decision. The subsequent debate over whether American participation in the Korean War was legally justified under U.S. law concerned the extent to which the president's constitutional powers (perhaps bolstered by congressional acquiescence or oblique endorsement) were legally adequate.[13] Viewed through the lens provided by *McCulloch* the "precedential" force of Korea ought to be located in the fact that President Truman and his advisors, willing though they were to engage in a bloody land war without any formal congressional authorization, were *not* willing to invoke international law obligations as providing legal authorization on its own. If Truman and Co. thought making that argument a legal bridge too far, it would be wise in that regard to follow their example.

13. The State Department vigorously defended the position that President Truman's constitutional authority as commander in chief empowered him to use military force to protect "the broad interests of Amercian foreign policy" including the continuing viability of the United Nations. *See Authority of the President to Repel the Attack in Korea*, Dep't State Bull. (July 31, 1950), at 173-77. For careful examinations of the constitutional debate, *see* Jane E. Stromseth, Rethinking War Powers: Congress, the President, and the United Nations, 81 GEO. L.J. 597, 621-40 (1993); CHARLES A. LOFGREN, GOVERNMENT FROM REFLECTION AND CHOICE 206–25 (1986).

3. Unspecified "Other" Sources of Legal Authority

The DOJ White Paper introduces its claims of authorization based on the president's "constitutional responsibility" and international law with language suggesting that it was not providing an exhaustive list of the sources of authority for the policy: "*In addition* to the authority arising from the AUMF, the President's use of force against al-Qa-ida and associated forces is lawful under *other principles* of U.S. and international law. . . ." The temptation to employ this sort of empty language was hardly new to the authors of the White Paper, and the classic argument why constitutional lawyers should discount it entirely in discussing presidential authority is Justice Jackson's concurrence in the Steel Seizure Case. "Loose and irresponsible use of adjectives colors all non-legal and much legal discussion of presidential powers. 'Inherent' powers, 'implied' powers, 'incidental' powers, 'plenary' powers, 'war' powers and 'emergency' powers are used, often interchangeably and without fixed or ascertainable meanings." Jackson readily conceded that such "nebulous claims" can serve as "a persuasive dialectical weapon in political controversy," but a central theme in his opinion was his insistence that vague and "claims of power" are of no value "in answering a constitutional question."[14]

Justice Jackson's rejection in *Youngstown* of the analytical value of vague and undefined language applies with equal force in any discussion of targeted killing. References to unidentified legal authorities, powers, or principles cannot be analyzed for their cogency: They do not invite intellectual engagement but

14. *Youngstown*, 343 U.S. at 646–47 (Jackson, J., concurring). Justice Jackson, it should be noted, also disavowed "the rigidity dictated by a doctrinaire textualism" that might render the president's powers as commander in chief so narrow that they would not achieve their constitutional ends. *Id.* at 640.

insinuate that the writer has reasons for his or her conclusions that the reader should take on faith. The atmospheric suggestion that there are additional legal bases for the targeted killing policy, uncoupled from any specification of what those bases may be, ought not to influence our thinking on the matter. The same is equally true, to be sure, of vague claims that the president *lacks* the legal authority to carry out the policy because of "the Constitution" or because it is "unthinkable" that he has that authority. Precisely because the strategic, political, and moral stakes are so high when the government of the United States lays claim to the authority to kill human beings, addressing the issue of legal authority demands hard-headed *legal* reasoning, not the deployment of the "dialectical weapons" of "political controversy."

4. The 2001 Authorization for the Use of Military Force

On September 18, 2001, Congress enacted the Authorization for the Use of Military Force.[15] Congress did not label the AUMF a declaration of war, and as a matter of parliamentary form it was a joint resolution rather than an act, but as we have already seen, neither fact is constitutionally significant. Congress may authorize military action to the extent that it deems appropriate by any statutory means it chooses. A joint resolution satisfies the constitutional requirements of bicameral passage and presentment to the president and thus has just as much the force of law as a statute packaged in the form of an "act of Congress." It is clear that Congress was authorized to use its war powers in response to 9/11, in light of several considerations: the level of

15. Pub. L. 107–40, 115 Stat. 224 (2001).

violence al-Qaeda had employed and which it was reasonable to believe the organization would repeat as opportunities became available; al-Qaeda's geopolitical goals which threatened not just individual Americans but the interests and security of the American political community; al-Qaeda's geographic location in places not subject to U.S. governance, making any kind of law enforcement response impossible; and the likelihood that the attacks had been aided or facilitated by the rogue Taliban regime in Afghanistan. Congress had the power to authorize hostilities, and it did so. Within its terms, therefore, the AUMF was an exercise of Congress's constitutionally authorized war powers to authorize the president in turn to use military force to respond to the attacks of 9/11 and to protect the Republic against any further such attacks.

Despite the Bush administration's preference for an extremely broad delegation of authority,[16] the text of the AUMF does not purport to issue the president a blank check to carry out whatever hostilities he deems necessary.[17] The Bush administration initially requested an authorization so broad as to come close to a

16. *See* David Abramowitz, *The President, the Congress, and Use of Force: Legal and Political Considerations in Authorizing Use of Force Against International Terrorism*, 43 Harv. Int'l L.J. 71, 74 n.8 (2002) (original White House proposals included "a request for the Congress to appropriate such sums as he deemed necessary to remain available until such funds were expended (transferring the appropriations power to the president); a request for the authority to waive any restriction on foreign assistance (such as Congressional mandates prohibiting assistance to state sponsors of terrorism, states responsible for gross violations of human rights, and states that were responsible for the proliferating of ballistic missiles and weapons of mass destruction) without notification or consultation with Congress; and restrictions on the provision of classified or sensitive law enforcement information to . . . relevant congressional committees")). *See also id.* ("the results of many [congressional] proposals were far narrower than those put forth initially by the President").

17. *Cf.* Hamdan v. Rumsfeld, 548 U.S. 557, 636 (2006) (Breyer, J., concurring) ("Congress has not issued the Executive a 'blank check'" to use military commissions in lieu of regular judicial venues).

general license to fight terror,[18] but as enacted, the language of the AUMF was considerably more precise, especially if one takes into account the urgency with which the administration was pressing Congress for the legislation.[19] Section 2(a) of the AUMF provides:

> That the President is authorized to use all necessary and appropriate force against those nations, organizations, or persons he determines planned, authorized, committed, or aided the terrorist attacks that occurred on September 11,

In one of his Cambodian sanctuaries opinions, Assistant Attorney General Rehnquist explained his conclusion that Congress may authorize limited hostilities and thus retain greater control over the president's discretion by drawing a contrast between a limited authorization and a declaration of war: "A declaration of war by Congress is in effect a blank check to the Executive to conduct military operations to bring about subjugation of the nation against whom war has been declared." *The President and the War Power: South Vietnam and the Cambodian Sanctuaries*, 1 Op. O.L.C. Supp. 321, 332 (1970). Even that rather unguarded statement notes one limitation on a declared war—the hostilities authorized must be directed to that "nation against whom war has been declared"—which suggests that Congress is not authorized to declare war, full stop, and empower the president to use military force against whomever the president thinks appropriate. It is unclear to me that Congress is constitutionally authorized to write the president blank checks in any issue involving the war powers, but in any event, the current AUMF is clearly not one.

18. The White House drafted a proposed joint resolution reading:

> That the President is authorized to use all necessary and appropriate force against those nations, organizations, or persons he determines planned, authorized, harbored [*sic*], committed, or aided in the planning or commission of the attacks against the United States that occurred on September 11, 2001, and to deter and pre-empt any future acts of terrorism or aggression against the United States.

Abramowitz, *supra* note 16, at 73 (quoting the administration draft). As Abramowitz indicates, if this draft had been enacted, the president would have possessed something very close to Assistant Attorney General Rehnquist's blank check.

19. Professor Matthew Waxman, whose expertise on the matter is both academic and practical (Waxman served in various high positions during the Bush administration), views the differences between the administration's proposed language and the enacted AUMF as cosmetic, and in particular he believes that the AUMF's language was not intended to be treated as placing specific, significant limitations on the president's discretion in fighting international terrorism. I disagree with Professor Waxman reluctantly, but whatever the particular intentions

> 2001, or harbored such organizations or persons, in order to prevent any future acts of international terrorism against the United States by such nations, organizations or persons.[20]

Section 2(a) defines the hostilities it authorizes in three ways. First, the section names the enemy. Traditionally, declarations of war and authorizations of more limited hostilities alike identified particular nations as the objects of the use of military force. Given that 9/11 was not perpetrated by a legitimate international sovereign and that it was "impossible [for Congress] to determine with any certainty exactly who would be implicated in the attacks,"[21] the AUMF does not name a specific foe, but at the same time it is not a blanket declaration of hostilities against anyone engaged in what the president might deem "international terrorism." Instead it specifically designates as the enemies against whom the president is authorized to use force "those nations, organizations, or persons" involved in 9/11 or sheltering those who were.[22] The White Paper's paraphrase of these legislatively identified enemies as "al-Qa'ida and its associated forces" reflects the practical definition of the AUMF definition worked out between the executive branch and the federal courts over the course of the ongoing Guantanamo habeas corpus litigation, and for current purposes we need not examine it in detail or determine whether it is

or expectations of the various political actors, it still seems to me highly important that what Congress passed and the president signed was a statutory text that fits Assistant Attorney General Rehnquist's description of a limited rather than a general authorization.

20. Pub. L. 107–40 § 2(a), 115 Stat. 224, 224, 50 U.S.C. § 5041 note (2001).

21. Abramowitz, *supra* note 16, at 74.

22. *See* Jennifer Daskal & Stephen I. Vladeck, *After the AUMF*, 5 HARV. NAT. SEC. L.J. 116, 116 (2014) ("Congress pointedly refused to declare a 'war on terrorism'" and instead limited its authorization to hostilities against "those responsible for the September 11 attacks, and those countries harboring the responsible parties").

entirely consistent with the AUMF.[23] As a panel of the D.C. Circuit has put it in the context of detention, "[t]he statutes authorizing the use of force and detention not only grant the government the power to craft a workable legal standard to identify individuals it can detain, but also cabin the application of these definitions."[24]

In crafting a definition of the enemies against whom it was approving the use of force, Congress was necessarily implying the existence of the class of those against whom it was *not* authorizing hostilities: "enumeration presupposes something not enumerated."[25] There is no reason to doubt that the D.C. Circuit panel was correct in its conclusion that the AUMF should be read to give the president broad discretion to determine who the designated enemies are, as a factual matter, in light of the statutory definition that limits the class of enemies to "those nations, organizations, or persons he determines" were involved in 9/11 or protecting those who were. But however broad and flexible the AUMF's definition of enemies, it is a definition, and one that necessarily limits the scope of the president's authority. At some point of factual disconnection or temporal remoteness from the events of 9/11, an organization or person simply would not be within the scope of the AUMF.

23. *See, e.g.*, Awad v. United States, 608 F.3d 1, 3 (D.C. Cir. 2010) ("Acting under the authority of the AUMF, the United States initiated a military campaign in Afghanistan against the Taliban regime and the al Qaeda forces it protected"), *cert. denied*, 131 S. Ct. 1814 (2011); al-Bihani v. Obama, 590 F.3d 866, 872 (D.C. Cir. 2010) (AUMF and the Military Commission Acts of 2006 and 2009 authorize military detention of "those who are part of forces associated with Al Qaeda or the Taliban or those who purposefully and materially support such forces in hostilities against U.S. Coalition partners"), *cert. denied*, 131 S. Ct. 1814 (2011). *See also* Hamdi v. Rumsfeld, 542 U.S. 507, 518 (2004) (plurality opinion of O'Connor, J.) ("There can be no doubt that individuals who fought against the United States in Afghanistan as part of the Taliban, an organization known to have supported the al Qaeda terrorist network responsible for those attacks, are individuals Congress sought to target in passing the AUMF.").

24. *Al-Bihani*, 590 F.3d at 872.

25. Gibbons v. Ogden, 22 U.S. (9 Wheat.) 1, 195 (1824).

Second, Congress announced the purpose for which it was authorizing the use of force: Section 2(a) authorizes the president to use military force "in order to prevent any future acts of international terrorism against the United States *by such nations, organizations or persons*." The AUMF provides the president with no statutory authority to use military force to protect the United States against future acts of international terrorism in general, but only for the specific purpose of preventing those defined by AUMF as enemies from engaging in future acts of international terrorism against the United States. The executive has no authority under the AUMF for other purposes, however sensible or even in some broad sense related to the purpose of the AUMF: Drone strikes to deter a repetition of 9/11 by a rogue state or organization with no connection to al-Qaeda or to prevent acts of international terrorism sparked by some wholly separate world conflict are not authorized by the statute. Once again, it would be appropriate to give section 2(a) a broad and flexible construction, but the legislation's definition of its purpose makes it impossible to treat the AUMF as authorization to suppress international terrorism generally.

Third, the AUMF authorizes the president "to use all necessary and appropriate force." The adjectives are reminiscent of the constitutional phrase "necessary and proper," which ever since *McCulloch v. Maryland* has been understood to grant Congress wide discretion in selecting the means for carrying into execution the powers that the Constitution provides it.[26] One might reasonably infer that the AUMF's similar language should be given a

26. 17 U.S. (4 Wheat.) 316 (1819). The echo in the AUMF of the Constitution's language is even clearer when one recalls the Supreme Court's traditional view that the constitutional amendments delegating to Congress power to enforce their provisions by "appropriate legislation" should be read to accord Congress the same broad discretion *McCulloch* found in the Necessary and Proper Clause. *See, e.g.,*

similar meaning, and thus provides the president with wide discretion in executing the powers that the AUMF provides him.[27]

Justice O'Connor's controlling opinion in *Hamdi v. Rumsfeld* is consistent with this inference. In *Hamdi*, the Court addressed the question whether the executive branch had the authority to detain a U.S. citizen believed by governmental officials to be a former member of a Taliban military unit; the executive intended to detain Hamdi indefinitely without charging him with a crime or permitting him an effective means of challenging the factual basis for his detention. The Court concluded that the AUMF's authorization of "all necessary and appropriate force" extended to the detention of enemy combatants during the course of the conflict.[28]

> The capture and detention of lawful combatants and the capture, detention, and trial of unlawful combatants, by "universal agreement and practice," are "important incident[s]

Katzenbach v. Morgan, 384 U.S. 641, 651 (1966) ("the *McCulloch v. Maryland* standard is the measure of what constitutes 'appropriate legislation'").

27. *See* Curtis A. Bradley & Jack L. Goldsmith, *Congressional Authorization and the War on Terrorism*, 118 HARV. L. REV. 2047, 2081 (2005) ("It seems unlikely that Congress, which views the Necessary and Proper Clause expansively, and has the most to gain from a broad interpretation of the clause, would have used the phrase 'necessary and appropriate' as a way to constrain presidential authority."). *Cf.* al-Bihani v. Obama, 619 F.3d 1, 25 n. 9 (D.C. Cir. 2010) (Kavanaugh, Cir. J., concurring in the denial of rehearing en banc) ("The 2001 AUMF is broader than certain other force authorizations that have limited the scope of authorized presidential action by regulating the method of force, controlling the use of military resources, setting limits on the timing, or dictating or limiting the choice of targets. For example, in November 1993, Congress authorized the president to use the U.S. Armed Forces in Somalia, but only through March 1994 and only for two specific, limited purposes—protecting 'United States personnel and bases' and 'securing open lines of communication for the free flow of supplies and relief operations.'") (citation omitted).

28. 542 U.S. 507 (2004). Justice Souter, speaking for himself and Justice Ginsburg, joined in the judgment. *See id.* at 553 (Souter, J., concurring in part, dissenting in part, and concurring in the judgment) ("the need to give practical effect to the conclusions of eight Members of the Court rejecting the Government's position calls for me to join with the plurality in ordering remand on terms closest to those I would impose").

> of war." The purpose of detention is to prevent captured individuals from returning to the field of battle and taking up arms once again.... In light of these principles, it is of no moment that the AUMF does not use specific language of detention. Because detention to prevent a combatant's return to the battlefield is a fundamental incident of waging war, in permitting the use of "necessary and appropriate force," Congress has clearly and unmistakably authorized detention in the narrow circumstances considered here.[29]

Justice O'Connor's reasoning is clear enough that we can generalize from the specific question ("the narrow circumstances considered here") of a citizen captured in a war zone and detained as an enemy combatant she addressed. The AUMF's language is not self-defining, but that does not make it impossible to construe: In authorizing the president to use "*all* necessary and appropriate [military] *force*," Congress must have intended him to have available a broad category of military measures but not just anything imaginable (the means must be "appropriate"), and military measures that are among the "incidents of waging war" accepted by universal agreement and practice are in that category.[30] The international law of war thus serves to inform the scope of authority conferred by the AUMF, although nothing in O'Connor's opinion necessarily entails the further conclusion

29. 542 U.S. at 518–19 (plurality opinion of O'Connor, J.) (citation omitted).

30. Justice Souter summarized this line of reasoning succinctly: "Because the Force Resolution authorizes the use of military force in acts of war by the United States, the argument goes, it is reasonably clear that the military and its Commander in Chief are authorized to deal with enemy belligerents according to the treaties and customs known collectively as the laws of war." *Id.* at 548 (Souter, J., concurring in part, dissenting in part, and concurring in the judgment). *See also* Bradley & Goldsmith, *supra* note 27, at 2081–82 ("Executive Branch practice in prior wars and the international laws of war inform the boundaries of a broad authorization of

that the law of war sets an outer limit to the measures the AUMF authorizes.[31]

Detention to prevent a combatant taking up arms again is, as Justice O'Connor pointed out, an established incident of waging war, and therefore under her view of the AUMF such detentions are authorized by the statute. It is equally clear that the law of war permits the use of deadly military force against enemy combatants—indeed, that would seem to be the core meaning of "war," as understood in our legal tradition at any rate. There is only one possible reading of the AUMF in that regard: It authorizes the use of military force, up to and including the killing

force such as the AUMF, and the phrase 'necessary and appropriate' can be viewed as encompassing these boundaries").

If the law of war were thought to authorize an action prohibited by a federal statute other than the AUMF, a question might arise whether the AUMF silently amended or repealed the earlier law. The answer to this highly theoretical question is no: The AUMF's language, with its reference to "appropriate" means, is a highly unlikely vehicle for modifying statutes such as the Uniform Code of Military Justice and the Torture Act in which Congress has imposed limitations on the behavior of U.S. military personnel. *See* U.S. CONST. art. I, § 8, cl. 14 ("The Congress shall have Power . . . To make Rules for the Government and Regulation of the land and naval Forces"); *Youngstown*, 343 U.S. at 644 (Jackson, J., concurring) (under the Government and Regulation Clause Congress "may to some unknown extent impinge upon even [presidential] command functions"). *See also* Nat'l Assoc. of Homebuilders v. Defenders of Wildlife, 551 U.S. 644, 664 n.8 (2007) ("Amendments by implication . . . are not favored") (citation omitted).

31. Various judges of the D.C. Circuit engaged in a dispute in the *al-Bihani* litigation over the role of international law in determining the scope of the authority the AUMF provides, including the possibility that the statutory grant of authority is limited to the actions permitted by the customary law of war. *See* the separate opinions filed when the court of appeals denied the appellant's petition for rehearing en banc, *al-Bihani v. Obama*, 619 F.3d 1 (D.C. Cir. 2010). The issue is irrelevant for our purposes: The targeted killing of al-Awlaki was a military measure involving the use of a weapon not currently prohibited by U.S. treaty or common international agreement, in contrast for example to chemical weapons. *See* Bradley & Goldsmith, *supra* note 27, at 2097 ("Although the laws of war inform the boundaries of what the AUMF authorizes, that simply means that as a general matter the AUMF authorizes no more than what the laws of war permit, not that it incorporates law-of-war prohibitions.").

of enemy combatants who are within the AUMF's broad limits. Even Justice Souter, who was skeptical about the AUMF's conferral of detention authority, agreed on this point. "Since the Force Resolution was adopted one week after the attacks of September 11, 2001, it naturally speaks with some generality, but its focus is clear, and that is on the use of military power. It is fairly read to authorize the use of armies and weapons, whether against other armies or individual terrorists."[32] This is not a blank check: It would be strange and unconvincing to read the general language of the AUMF, with its reference to "appropriate" uses of force, to amend or repeal existing U.S. law that prohibits the killing of noncombatants and combatants who surrender.[33] But the AUMF unquestionably authorizes the president, through the U.S. armed forces, to employ deadly force against enemy combatants.

5. The 2001 AUMF Authorized the al-Awlaki Targeted Killing by Drone

The AUMF authorizes the president to direct the U.S. armed forces to use deadly force against the enemies Congress has identified. The Supreme Court has authoritatively construed the AUMF's reference to "necessary and appropriate force" to include military actions that are traditional incidents of waging war. The use of any deadly weapon not itself clearly banned by international law against legitimate targets—basically enemy combatants—is one of those incidents. Drones are not (currently) the subject of a widespread, agreed-upon

32. *Hamdi*, 542 U.S. at 547 (Souter, J., concurring in part, dissenting in part, and concurring in the judgment).

33. More generally, for our purposes I think we may assume that the AUMF preserves the limitations on whom the U.S. military may kill, and under what circumstances, that U.S. statutes and military regulations otherwise impose.

international ban.[34] The targeted killing by drone of members of al-Qaeda who are enemy combatants is thus within the necessary and appropriate uses of military force that Congress has authorized. At least as long as the targeted killing policy remains within the limits of the traditional law of war, and the AUMF is in effect, as a matter of statutory construction the use of drones in execution of the policy is clearly within the president's lawful authority.[35]

This proposition needs no direct elaboration, but there are a few side issues. The DOJ White Paper defines quite narrowly the circumstances that it addresses and the "conditions" under which it concludes that a targeted killing would be lawful: The analysis is limited to the targeting of "a U.S citizen who is a senior operational leader of al-Qa'ida or an associated force of al-Qa'ida—that is, an al-Qa'ida leader actively engaged in planning operations to kill Americans," and does not purport to "to determine the minimum requirements necessary to render such an operation lawful, nor [to] assess what might be required to render a lethal operation against a U.S. citizen lawful in other circumstances." One of the White Paper's "conditions" involves conformity to the law of war limits on targeting individuals. In the next paragraph I explain my conclusion that the AUMF should be read to require

34. *See, e.g.*, Interview with Peter Maurer, President of the International Committee of the Red Cross (May 5, 2013), https://www.icrc.org/eng/resources/documents/interview/2013/05-10-drone-weapons-ihl.htm.

"Under international humanitarian law—the rules of war, i.e. the set of laws governing armed conflicts—drones are not expressly prohibited, nor are they considered to be inherently indiscriminate or perfidious. In this respect, they are no different from weapons launched from manned aircraft such as helicopters or other combat aircraft."

35. *See Hamdi*, 542 U.S. at 547 (Souter, J., concurring in part, dissenting in part, and concurring in the judgment) (the AUMF "is fairly read to authorize the use of armies and weapons . . . against . . . individual terrorists"); al-Bihani v. Obama, 619 F.3d at 24 (Kavanaugh, Cir.J., concurring in the denial of rehearing en banc) ("The AUMF affords the President broad discretion with respect to methods of force, use of military resources, timing, and choice of targets").

such conformity; if I am correct, then this "condition" *is* a one of the minimum requirements necessary to render lawful the targeted killing of *anyone* pursuant to the AUMF. Neither of the other two White Paper "conditions" (see the footnote) states a qualification on the executive's legal authority to order the use of deadly force: There is no reason to read into the AUMF any limitations about who must make a particular decision, or whether the enemy combatant poses any particular level of threat, or whether it would be possible to capture rather than kill him.[36]

Neither the OLC Memo nor the DOJ White Paper nor any official statement about the targeted killing policy of which I am aware claims (or admits) that the policy has been extended to the deliberate targeting of persons who may not be targeted under the law of war. Justice O'Connor's controlling opinion in *Hamdi v. Rumsfeld* establishes that military action within the limits of the law of war is authorized by the AUMF, but as I observed earlier, I would not read her to give a definitive answer to the general question whether the AUMF authorizes other actions not within those limits. However, we may infer the better answer with respect to the specific question of who may be targeted: The AUMF should not be construed to authorize the targeting of individuals who are immune from deliberate attack under the law of war. *Hamdi* addressed the question of what the president is authorized to do to an individual with respect to indefinite detention and answered that as an enemy combatant the

36. DOJ White Paper, *supra* note 6, at 1. The White Paper concluded that the following conditions "would be sufficient to make lawful" such an operation: "(1) an informed, high-level official of the U.S. government has determined that the targeted individual poses an imminent threat of violent attack against the United States; (2) capture is infeasible, and the United States continues to monitor whether capture becomes feasible; and (3) and the operation is conducted in a manner consistent with the four fundamental principles of the laws of war governing the use of force." *Id.* at 16. The White Paper does not assert that these conditions are the minimal legal requirement but only that meeting them assures the legality of a targeted killing.

petitioner was subject to detention to prevent him from returning to the battlefield; O'Connor looked to the law of war both for the authorization she found in the AUMF and for its limitations (not for punishment, not to extend beyond the end of hostilities). The question of what the president may do in the way of ordering the military to use deadly force against an individual is similar in logical form. I see no reason why we should not answer it using O'Connor's approach and construe the AUMF with a view to the law of war, which authorizes the deliberate killing of enemy combatants but not of noncombatants. This conclusion is supported by the old canon of statutory construction that "'an act of congress ought never to be construed to violate the law of nations if any other possible construction remains.'"[37]

I reach the opposite conclusion with respect to the argument that the AUMF necessarily incorporates international law rules about respecting the sovereignty of independent states. On its face the AUMF imposes no geographic restrictions on the hostilities it authorizes, and indeed it could not have done so: When Congress enacted the statute it understood that it was authorizing hostilities against enemies whose identities and locations were uncertain, some of whom would probably not be agents of an internationally recognized government or exercise even de facto control over any particular geographic region. The language of the AUMF ("against those nations, organizations, or persons" involved in 9/11) is clear evidence of Congress's intention to authorize the use of force against nonstate entities including, conceivably, those that could *only* be located within territory under the sovereign jurisdiction of a state or states against which Congress was not authorizing hostilities.[38]

37. Murray v. Schooner Charming Betsy, 6 U.S. (2 Cranch) 64, 118 (1804) (Marshall, C.J.).

38. The AUMF authorizes the use of force "against those nations, organizations, or persons" the president determines were involved in 9/11 (centrally, al-Qaeda),

A thought experiment: Suppose there had been no regime harboring al-Qaeda that was exercising at least de facto governmental authority over a state or coherent geographic area (arguably the case in 2011), and that the president was advised that international law forbade the use of military force against any of the states in which al-Qaeda members were known to be operating militarily.[39] Would it have been necessary or appropriate for the president to conclude that the AUMF could not be executed? Such a conclusion, I think, would nullify the clear purpose of the statute to empower the president to use military force, within the AUMF's limits, "to prevent any future acts of international terrorism against the United States" by the enemies designated by the statutory language. If international law rules concerning the inviolability of national sovereignty were to prevent the president's ability to pursue the purpose of the AUMF, it would render the AUMF self-defeating to read those rules into it. The reader will recall the *McCulloch* principle that unless the language of the Constitution demands otherwise we should construe the powers of the federal government as adequate to meet constitutional purposes. Given the fundamental purpose of the

"*or harbored* such organizations or persons." In *al-Bihani v. Obama*, a panel of the D.C. Circuit construed the statutory term "harbored" by treating the Taliban regime that controlled Afghanistan until its overthrow in late 2001 as the paradigm example of such an organization. The Taliban, it reasoned, "harbored" al-Qaeda by providing al-Qaeda a "safe haven in Afghanistan"; the 55th Arab Brigade, the paramilitary organization with which the appellant served, therefore fell "within the AUMF's wide ambit as an organization that harbored Al Qaeda" because "the actual and foreseeable result of the 55th's defense of the Taliban was the maintenance of Al Qaeda's safe haven." 590 F.3d at 874. This seems a reasonable interpretation of the statute. In contrast, it is easy to imagine an internationally recognized government that had no intention of providing al-Qaeda with a safe haven and took no actions that would foreseeably create such a safe haven, and yet was unable to oust al-Qaeda from its territory.

39. The first of these stipulations was, of course, counterfactual in the fall of 2001, and the second raises issues about international law that are in dispute. *See, e.g.*, DOJ White Paper, *supra* note 6, at 3–5 (describing international law issues related to territorial sovereignty).

war powers—collectively they are the "power to wage war successfully"—I do not think there is any constitutional difficulty with construing the AUMF to authorize the use of military force notwithstanding international-law limitations grounded in the sovereignty of other countries.[40]

6. The President Did Not Have Extrastatutory Authority under Article II to Order the al-Awlaki Killing

The DOJ White Paper treats Article II of the Constitution and the AUMF as complementary and equally available sources of legal authority for the president's decisions in the war on terror, including the decision to engage in targeted killing: "It was on these bases that the United States responded to the attacks of September 11, 2001, and '[t]hese domestic and international legal authorities continue to this day.'" On this point, I believe that the White Paper has fallen into error: The enactment of the AUMF significantly limits the president's legitimate authority to

40. As I discuss below, the conclusion that Congress or the president have the constitutional authority to do X does not entail or even suggest the further conclusion that they should do X. While I accept the common U.S. "dualistic" view of international law as distinct from domestic U.S. law, I also accept the proposition, which goes back to the early Republic, that the United States is subject to the law of nations and that when the U.S. government violates a clear international law norm it is acting lawlessly *from the perspective of that system of legal norms*. For a recent Supreme Court decision that (I believe) accepts this position, *see* Medellin v. Texas, 552 U.S. 491 (2008). In some circumstances, all three branches of the federal government may have the constitutional authority, and even the constitutional duty, to disregard international law constraints, but it is no light matter for the government of a Republic constituted by law to violate laws the authority of which it ought to respect. Even for those who do not accept the normative force of international law, the geopolitical consequences of American violations of customary international practices can be costly.

invoke independent Article II authority in pursuing the hostilities that the AUMF addresses.

Justice Jackson's *Youngstown* opinion provides the proper baseline for analysis: "Presidential powers are not fixed but fluctuate, depending upon their disjunction or conjunction with those of Congress."[41] The powers that the president may exercise in the absence of legislation, when "congressional inertia, indifference or quiescence may . . . enable, if not invite, measures on independent presidential responsibility," are not necessarily, indeed not ordinarily available when the president wishes to act contrary to the enacted will of Congress. In the latter situation, the president "can rely only upon his own constitutional powers minus any constitutional powers of Congress over the matter."

> Courts can sustain exclusive Presidential control in such a case only by disabling the Congress from acting upon the subject. Presidential claim to a power at once so conclusive and preclusive must be scrutinized with caution, for what is at stake is the equilibrium established by our constitutional system.

As a formal matter, the constitutional text distinguishes the roles of Congress, which is constitutionally authorized to make law, and the president, who is constitutionally obliged to "take Care that the Laws be faithfully executed."[42] From a political science or legal realist perspective the reality is a good deal more complicated, and much of the vast web of federal legal norms "on the

41. 343 U.S. at 635 (Jackson, J., concurring).

42. *See, e.g.*, *Youngstown*, 343 U.S. 579, 587–88 (Black, J.) (1952) ("In the framework of our Constitution, the President's power to see that the laws are faithfully executed refutes the idea that he is to be a lawmaker. . . . And the Constitution is neither silent nor equivocal about who shall make laws which the President is to execute. The first section of the first article says that 'All legislative Powers herein granted shall be vested in a Congress of the United States.'").

ground" are the product to varying degrees of policy decisions by executive and administrative officials.[43] Nonetheless the political system continues to function on the basis of the text-based asymmetry between legislative and executive authority: Congress's legislative jurisdiction is broad in itself and extends to the enactment of legislation *Congress* deems "necessary and proper" for the execution of the president's powers, while the president is under at the least a presumptive duty to execute Congress's laws, including those regulating his exercise of his own constitutional powers.

A presidential power that is so "conclusive and preclusive" that it authorizes the president to violate an act of Congress is therefore contrary to the ordinary working of the system. It is tantamount to the conclusion that Congress cannot use law to regulate the use of a constitutionally authorized power in the teeth of Article I, section 8's delegation to Congress of the authority to do so with respect to "*all* . . . Powers vested by this Constitution in the Government of the United States, or in *any* Department or *Officer* thereof." "Courts can sustain exclusive Presidential control in such a case only by disabling the Congress from acting upon the subject."[44]

Jackson did not deny the existence of "conclusive and exclusive" presidential powers, but it is clear from his opinion that he thought them few and far between. His elaborate discussion of the commander in chief power suggests that what Jackson calls the "command functions" of the president involve an area of exclusive

43. The power of the president, with the advice and consent of the Senate, to make domestic law through a self-executing treaty adds another level of complexity, although such treaties constitute only a small part of federal law.

44. *Id.* at 637–38 (Jackson, J., concurring). In one respect, I think Justice Jackson has slightly overstated his conclusion. Even in an area of "conclusive and preclusive" presidential authority, Congress would still have the Necessary and Proper Clause power to legislate in support of the president's exercise of authority, e.g., the exclusivity of the veto power does not preclude Congress from providing the president with resources or personnel for gathering information on veto applications, or

executive control, but most of his argument is crafted to narrow and cabin in the legitimate claims of any such exclusive command authority: He expressly rejected the idea that the Commander in Chief Clause accords the president a general power "to do anything, anywhere, that can be done with an army or navy," and noted that through its enumerated power "to make rules for the 'Government and Regulation of land and naval forces,'" Congress may "to some unknown extent impinge upon even command functions."[45]

The modern Supreme Court views Justice Jackson's "familiar tripartite scheme" as "the accepted framework for evaluating executive action" at least in matters touching on foreign affairs; that framework in turn operationalizes Jackson's basic principle that presidential authority is affected by, and ordinarily limited by, Congress's exercise of its powers.[46] With respect to matters over which the Constitution has delegated substantive lawmaking authority to Congress, and where Congress has spoken, Congress is the final policy maker, and the president's Article II

insuring that vetoes are respected in a timely fashion. *See* U.S. CONST. art. I, § 8, cl. 18 ("The Congress shall have Power . . . To make all Laws which shall be necessary and proper for carrying into Execution the foregoing Powers, and *all other Powers vested by this Constitution in the Government of the United States, or in any Department or Officer thereof.*") (emphasis added). This tweaking of Jackson's language has no effect on the fluctuation in executive power relative to Congress's exercise of the specific enumerated powers the Constitution delegates to the legislative branch.

45. *Youngstown*, 343 U.S. at 644–45, 641–42, 644 (Jackson, J., concurring). *See also id.* at 645–46 (the president's "command power is not such an absolute as might be implied from that office in a militaristic system but is subject to limitations consistent with a constitutional Republic whose law and policy-making branch is a representative Congress"). Jackson's disparaging references to the Korean War as Truman's "foreign venture" and "Korean enterprise," *id.* at 642, 643, hardly suggest an expansive view of unilateral presidential authority to initiate hostilities.

46. *See* Zivotofsky v. Kerry, 135 S. Ct. 2076, 2083 (2015) ("In considering claims of Presidential power this Court refers to Justice Jackson's familiar tripartite framework from *Youngstown*"). *See also* Dames & Moore v. Regan, 453 U.S. 654, 668 (1981) ("Justice Jackson's concurring opinion elaborated in a general way the consequences of different types of interaction between the two democratic branches in assessing Presidential authority to act in any given case.").

"executive Power" is the authority to "direct that [the] congressional policy be executed in a manner prescribed by Congress," and not to substitute "a presidential policy [to] be executed in a manner prescribed by the President."[47] The ordinary constitutional rule, therefore, is that the president has no lawful authority to act in defiance of a validly enacted statute.[48]

This rule, if it applies, is inconsistent with the DOJ White Paper's assertion that the president's authority under Article II to respond to 9/11 "continue[s] to this day," apparently unaffected by Congress's enactment of legislation, the AUMF, that directly addresses what the president may do. The AUMF is, we saw earlier, an exercise of Congress's enumerated, substantive powers, powers that there is no textual reason to treat as other than plenary. By defining the executive actions it authorizes, the AUMF limits the scope of those actions. Following Justice Jackson's "tripartite scheme" of analysis, the president would be legally entitled to ignore the limitations only if it would be legally proper to

47. *Youngstown*, 343 U.S. at 588 (Black, J.).

48. There is judicial and academic support for an absolute rule that "no presidential authority to act *contra legem* exists." Henry Paul Monaghan, *The Protective Power of the President*, 93 COLUM. L. REV. 1, 11 (1993). *See, e.g.*, Kendall v. United States ex rel. Stokes, 37 U.S. (12 Pet.) 524, 613 (1838) ("vesting in the President a dispensing power . . . has no countenance for its support in any part of the constitution"). As I indicate in the text, I believe that this formulation of the rule is too stark. Justice Jackson's "command functions" within the military and the conduct of diplomacy seem to me areas where the president's discretion is not subject to congressional control, and I would also think the president was entitled to disregard a statute that violated a separation of powers principle established by the Supreme Court such as the prohibition on legislative vetoes. But these are exceptions, however important in practice, to a broad, general rule.

No one claims that Congress may control the president's prerogative powers of the veto, pardons, and appointment, although of course the two houses may override a veto by supermajority vote and Congress may vest the appointment of inferior officers "in the Courts of Law, or the Heads of Departments." *See also Civil-Service Commission*, 13 Op. Att'y Gen. 516 (1871) (Congress may impose statutory qualifications for inferior officers as long as the qualifications leave adequate scope for the exercise of "judgment and will").

disable Congress from imposing its policy choices with respect to the issue in question, the scope of lawful hostilities—but this is the very core of the declaration of war power. As *Little v. Barreme* and the Steel Seizure Case indicate, the Commander in Chief Clause does not empower the president to act in ways inconsistent with the exercise of a plenary Article I power. (Four of the majority Justices in *Youngstown*—Jackson, Frankfurter, Burton, and Clark—thought decisive the existence of relevant legislation inconsistent with Truman's seizure; Justice Black, who thought that Truman simply lacked the power he had exercised regardless of what Congress had done, unquestionably agreed that on any issue within Congress's legislative jurisdiction the president is bound by Congress's actions.) It follows that however broad the president's unilateral Article II authority to respond to 9/11 was in Phase Two, before the enactment of the AUMF, once that statute was enacted the president was obliged to respect the definitional limitations it places on his authority. The Supreme Court indicated as much in the *Hamdan* military-commission case, decided in 2006: "Whether or not the president has independent power, absent congressional authorization, to convene military commissions, he may not disregard limitations that Congress has, in proper exercise of its own war powers, placed on his powers. *See Youngstown Sheet & Tube Co. v. Sawyer* (Jackson, J., concurring)."[49]

The *Hamdan* principle has a limit: It does not apply if Article II not only gives the president the power to act but is the rare example of what Justice Jackson termed a "conclusive and preclusive"

49. Hamdan v. Rumsfeld, 548 U.S. 557, 593 n.23 (2006) (citation abbreviated). This footnote is in the part of Justice Stevens's opinion in *Hamdan* that commanded a majority of the Justices. Stevens's inclusion of the word "proper" makes the assertion a tautology on its face, but it is an informative tautology once one recalls the very broad scope that a *McCulloch* approach to constitutional thought gives to Congress's war-related powers, and the fundamental asymmetry between legislative and executive power that Justice Jackson recognized in *Youngstown*.

presidential power that can "disabl[e] the Congress from acting upon the subject" in ways contrary to the president's choices.[50] In that case, and in that case only, it would not be a *proper* exercise of Congress's war powers for the legislative branch to limit the discretion of the executive. The obvious and only possible Article II power that might enable the president to escape the limits of the AUMF is that provided by the Commander in Chief Clause, and there are those who read the clause to confer such an illimitable power on the executive. We have already seen one fatal flaw in this view of Article II: It implicitly treats the Declaration of War Clause as less than plenary, uniquely so among Article I's unqualified grants of power, and does so without textual warrant and in the teeth of judicial commentary from Chief Justice Marshall to Chief Justice Rehnquist. Another flaw that should weigh heavily with many is that it seems highly implausible to interpret the Constitution, as a matter of original meaning, to grant the president the sort of illimitable power to conduct hostilities that characterized eighteenth-century monarchs:[51] Article I, section 8 assigns to Congress many of the military powers that had belonged to the king, a decision that in the founding era was most easily understood as the deliberate subordination of the executive to legislative control on those issues, including the initiation of hostilities.[52] Finally, the illimitable power viewpoint disrupts "the equilibrium established by our constitutional

50. *Youngstown*, 343 U.S. at 637–38 (Jackson, J., concurring).

51. *See id.* at 641 (Jackson, J., concurring) ("The example of such unlimited executive power that must have most impressed the forefathers was the prerogative exercised by George III, and the description of its evils in the Declaration of Independence leads me to doubt that they were creating their new Executive in his image.").

52. *Compare* 2 WILLIAM BLACKSTONE, COMMENTARIES ON THE LAWS OF ENGLAND *239–40 (1766) *with* FEDERALIST No. 69 (Alexander Hamilton), at 464–65 (Jacob E. Cooke ed., 1961) ("The President is to be [commander in chief] of the

system"[53] and does so in the absence of any exigent legal or political reason. It is possible to recognize robust presidential authority to use the armed forces to protect the Republic without upending the usual rule about the asymmetry between legislative and executive authority and treating Congress's power to determine which hostilities to authorize as a legal nullity that can be ignored by the president at will.

To summarize: Because the AUMF is the valid exercise of a plenary Article I power, the president is legally obligated to respect the limitations it places on his military decisions by defining which hostilities it is authorizing. The authority to make such decisions that the president derives directly from Article II is subject to regulation by Congress, which outside the specific command functions of the presidency "exercises a power of precedence over, [though] not exclusion of, Executive authority."[54] The president is bound by the limitations of the AUMF and may not invoke Article II to avoid those limitations.[55] This does not mean, to be sure, that the AUMF disables the president from

army and navy of the United States. In this respect his authority would be nominally the same with that of the king of Great Britain, but in substance much inferior to it. It would amount to nothing more than the supreme command and direction of the military and naval forces, as first General and admiral of the Confederacy; while that of the British king extends to the declaring of war and to the raising and regulating of fleets and armies, all which, by the Constitution under consideration, would appertain to the legislature."). The founders were probably more divided on the scope of the commander in chief power than partisans on either side of the modern debate concede, but it is hard to believe that many of them thought Article II gave the president a blank check to override Congress on everything having to do with the military, and especially Congress's view on when to initiate hostilities.

53. *Youngstown*, 343 U.S. at 638 (Jackson, J., concurring).

54. *Loving*, 517 U.S. 748, 767 (1996).

55. Unlikely or absurd hypotheticals do not advance constitutional discussion. The president and his advisors can and should consider the possibility that a particular application of the AUMF's limits is unconstitutional as applied when they confront a genuine conflict between military necessity and the statute's broad and pliable terms.

taking military action, within the limits of his unilateral Article II authority, against threats that are genuinely distinct from those presented by al-Qaeda, the Taliban and their allies. But with respect to the perpertrators of 9/11 Congress has defined (very broadly) the lawful scope and purpose of any U.S. military operations.

C. THE KILLING OF AL-AWLAKI DID NOT VIOLATE A CONSTITUTIONAL PROHIBITION

This book argues in Chapter Three that military action which is constitutionally authorized by the war powers of Congress and the president is not subject to constitutional prohibitions such as the Fifth Amendment. In his May 2013 speech, President Obama appeared to take the diametrically opposite position: "I do not believe it would be constitutional for the government to target and kill any U.S. citizen—with a drone, or a shotgun—without due process."[56] In this section I shall assume my view is erroneous and that the president and the Justice Department were correct to assume that U.S. citizen-targets abroad are entitled to the protection of due process.

Neither the OLC Memo nor the DOJ White Paper asserts as clear and unequivocal a position as the president's; they present their due process analysis, at least in form, as a discussion *arguendo*. "The Department assumes that the rights afforded by [the] Fifth Amendment's Due Process Clause, as well as the Fourth Amendment, attach to a U.S. citizen even while he is abroad."[57] The

56. Http://www.aclu.org/files/assets/obama-speech.pdf.

57. DOJ White Paper, *supra* note 6, at 5. The OLC Memo was seemingly closer to the president. *See* OLC Memo, *supra* note 7, at 38 (the amendments "likely protect [al-Awlaki as a citizen] in some respects even while he is abroad").

White Paper addresses its due process analysis, furthermore, to the precise policy that it describes, "a lethal operation against a U.S. citizen who is a senior operational leader of al-Qa'ia or its associated forces . . . and who himself poses an imminent threat of violent attack against the United States."[58] In addition, the administration's legal discussions have been limited to *procedural* due process, the constitutional requirement that government follow adequately fair procedures; there is no evidence that the executive's lawyers have considered any issues about targeted killing by drone that might be raised by what constitutional lawyers call *substantive* due process.

In discussing procedural due process, the OLC Memo and the DOJ White Paper follow Justice O'Connor's opinion in *Hamdi v. Rumsfeld* and adopt "the traditional due process balancing analysis of *Mathews v. Eldridge*" to "suppl[y] the framework for assessing the process due a U.S. citizen who is a senior operational leader of an enemy force planning violent attacks against Americans before he is subjected to lethal targeting."[59] The *Mathews* framework is a form of cost-benefit analysis, involving a comparison of the relative weights to be ascribed to the individual's interest in not being harmed and to the government's interest in taking the proposed action, and also requires consideration of the risk of government error and the value of any particular form of procedure. As a matter of first principles, one might question whether this mix of concerns is susceptible to the sort of quasi-algorithmic analysis *Mathews* requires, but the decision is positive law, and for present purposes we must accept it as such.

58. DOJ White Paper, *supra* note 6, at 2.

59. DOJ White Paper, *supra* note 6, at 2, 6 (citing *Hamdi*, 542 U.S. at 529 (plurality opinion) (citing Mathews v. Eldridge, 424 U.S. 319, 335 (1976)). *See* OLC Memo, *supra* note 7, at 39.

The OLC Memo and the White Paper admitted that a targeted individual has a "'uniquely compelling'" interest "in avoiding the erroneous deprivation of his life." At the same time, both documents concluded that the administration's policy satisfies the demands of due process: "[T]he weight of the government's interest in protecting its citizens from an imminent attack [is] such that the Constitution would not require the government to provide *further process* to such a U.S. citizen before using lethal force."[60] The policy process that the OLC Memo and the White Paper consider constitutionally adequate under the circumstances is supplied by the first two "conditions" that the policy stipulates for the targeted killings: a determination by "an informed, high-level official of the U.S. government" that the targeted citizen "poses an imminent threat of violent attack against the United States"; and a determination that "capture is infeasible" while the government "continues to monitor whether capture becomes feasible."[61]

The Justice Department's due process discussions have obvious, if superficial attractions: They appear to require deliberation rather than arbitrary decision, and, above all, they seem to avoid the uncomfortable conclusion that the U.S. government can kill a U.S. citizen without affording him due process. From the standpoint of the administration they have the additional virtue of upholding the constitutionality of the administration's targeted killing policy as it has been publicly described. I would agree that the described requirements and procedures are reasonable in some general sense. But as a piece of constitutional analysis

60. DOJ White Paper, *supra* note 6, at 6 (citation omitted and emphasis added). For similar reasons, the OLC Memo and the White Paper conclude that the targeted killings it discusses would not violate the Fourth Amendment. *See id.* at 9–10; OLC Memo, *supra* note 7, at 41.

61. DOJ White Paper, *supra* note 6, at 16. Presumably the continuing monitoring would allow and require the killing to be called off if capture became feasible. *See id.* at 9 (stating, in discussing the Fourth Amendment, that "those carrying out

meant to rest on *Hamdi v. Rumsfeld*, the due process discussion in the OLC Memo and the White Paper discussion is unpersuasive. To see why, recall that the due process issue before the Court in *Hamdi* was what procedures, if any, the executive branch was required to afford a U.S. citizen it was detaining as an enemy combatant who claimed that the government was simply in error about his pre-detention activities. In the controlling opinion, Justice O'Connor summarized the essential requirements of procedural due process.

> "An essential principle of due process is that a deprivation of life, liberty, or property 'be preceded by notice and opportunity for hearing appropriate to the nature of the case.'" "[D]ue process requires a 'neutral and detached judge in the first instance.'" "For more than a century the central meaning of procedural due process has been clear: 'Parties whose rights are to be affected are entitled to be heard; and in order that they may enjoy that right they must first be notified.' It is equally fundamental that the right to notice and an opportunity to be heard 'must be granted at a meaningful time and in a meaningful manner.'" These essential constitutional promises may not be eroded.

In light of this understanding of the "central meaning" of procedural due process, O'Connor wrote that "[w]e therefore hold that a

the operation would carry out the operation only if capture were infeasible"). The White Paper's third "condition," that the attack be carried out in accordance with "the four fundamental principles of the laws of war governing the use of force," *id.* at 16, has no obvious relationship to *procedural* due process, the constitutional limitation applied by *Mathews* and *Hamdi*. If those international law principles were relevant to the constitutional analysis of the targeted killing policy, adherence to them would be a factor in determining whether the government had violated the limitations imposed by *substantive* due process. *See* County of Sacramento v. Lewis, 523 U.S. 833, 846 (1998) (substantive due process prohibits "executive abuse of power . . . which shocks the conscience"). *See* the discussion in Chapter Five.

citizen-detainee seeking to challenge his classification as an enemy combatant must receive notice of the factual basis for his classification, and a fair opportunity to rebut the Government's factual assertions before a neutral decisionmaker."[62] In other words, the process by which the government decides it wishes to proceed against the individual (in Hamdi's case to detain him indefinitely) must be separate from the process by which the government evaluates the factual basis on which the first process has reached its conclusion, and the second process must take into account any claim by the individual that the first process was factually erroneous.

The internal executive branch deliberations that the OLC Memo and the White Paper appear to outline are light years away from what Justice O'Connor describes as the "essential constitutional promises" of due process. Of course the administration's lawyers did not propose that potential targets be given notice of the government's possible interest in killing them. Of course they did not contemplate, much less require, that a targeted individual be heard at any time or in any manner as to why the government is mistaken about his identity or activities. Of course they did not provide for a neutral and detached decision maker to resolve any factual uncertainty: The ultimate decision maker in the targeted killing program is the commander in chief himself, who neither can nor should be neutral or detached, and even if individual decisions are delegated to his subordinates, they too would be officials in the executive branch, the very branch of government that wishes to deprive the target of his life.[63] "'[O]ne is entitled as a matter of due process of law to an adjudicator who is not in a

62. *Hamdi*, 542 U.S. at 533 (plurality opinion) (citations and parentheses omitted).

63. *Id.* at 537–38 (plurality opinion). The government argued in *Hamdi* that the petitioner had an opportunity to contest its factual beliefs about him when he was screened and interrogated by executive branch or military officials. Justice

situation which would offer a possible temptation to the average man as a judge . . . which might lead him not to hold the balance nice, clear and true.' That even purportedly fair adjudicators 'are disqualified by their interest in the controversy to be decided is, of course, the general rule.'" None of these commonplace, settled elements of fair constitutional procedure play any part in the decision-making procedure the OLC Memo and the White Paper outline. The "process" that the administration concluded sufficient to meet constitutional requirements bears no resemblance to the requirements that *Hamdi* found minimally necessary.[64]

The Obama administration's processes, however reasonable they may be as a method of decision making in other respects, are incompatible with the *due* process Supreme Court precedent appears to require. The administration's response—and this is put harshly but I think accurately—has been to define its processes as constitutional by fiat, ignoring the plain teaching of the Court and thereby watering down the meaning of due process in the name of necessity. Since we can't in fact satisfy *any* of the "essential constitutional promises" of due process, the implicit reasoning goes, we should create other criteria that are faintly reminiscent of those promises and treat these as due process in

O'Connor's response was brisk: "An interrogation by one's captor, however effective an intelligence-gathering tool, hardly constitutes a constitutionally adequate fact-finding before a neutral decisionmaker. . . . Plainly, the 'process' Hamdi has received is not that to which he is entitled under the Due Process Clause." *Id.*

64. The judicial decisions holding that the due process is satisfied despite the absence of a pre-deprivation hearing "[w]hen the State must act quickly or predeprivation process is impracticable, and meaningful postdeprivation process is available" are not germane. *See* Bowlby v. City of Aberdeen, 681 F.3d 215, 220 n.1 (5th Cir. 2012) (citing, *inter alia*, Hudson v. Palmer, 468 U.S. 517 (1984)). Those cases address situations in which the deprivation of a protected liberty or property interest "is occasioned by a random, unauthorized act by a state employee, rather than by an established state procedure," *Hudson*, 468 U.S. at 532, and patently have no application to a decision to deprive someone of life taken after great deliberation by high officers in the U.S. government.

the context of targeted killing. This "argument" is, again, implicit since neither the OLC Memo nor the White Paper actually acknowledges the contradiction between the administration's views and the precedents. It saves the supposed principle that the government must always give a citizen "due process" before killing him . . . at the expense of evacuating that phrase of its traditional and settled content.

If the requirements set out in the OLC Memo and the DOJ White Paper provided a functional parallel to the demands of traditional due process, the argument that they should be treated as due process in this unique context might have some purchase, but I do not see that there is a genuine parallel in terms of function. The administration's procedures *in no sense* provide the person targeted with any voice, even a notional one, and *in no sense* do they place the decision to allow the killing in the hands of a neutral decision maker. They involve the executive branch talking to itself, and deciding whether it believes that its reasons for the killing it is contemplating are convincing. Better by far, I think, simply to concede that the administration's procedures, however desirable as a matter of good government practice, do not and cannot afford a potential target even the most minimal process that is due under the most relaxed reading of the Fifth Amendment. The Due Process Clause imagined in the OLC Memo and the White Paper is a chimera, and a delusive one at that. If the real Due Process Clause applies to the targeted killing policy, that policy is unconstitutional as a denial of procedural due process.[65]

* * *

65. In very helpful private conversations, Professors Philip Bobbitt and Ingrid Wuerth have argued strongly that the proper articulation of my basic claim is not that the Due Process Clause does not apply, but that the adoption of the AUMF supplies all the process that is due. This articulation of the position that the Due Process Clause does not prohibit targeted killings authorized by the AUMF may be more persuasive to some who find my starker formulation too troubling. Chief Justice

One theme of this book is that we should not place over-much reliance on constitutional law. Asking and answering the question whether targeted killing policy and practice are constitutional is a necessary but not sufficient step in determining whether that policy and practice should be continued, modified, or terminated. What the current administration says, and doesn't say, about due process illustrates the danger in allowing constitutional law, and constitutional lawyers, to play too large a role in our thinking about military affairs. From the standpoint of constitutional law, the preservation of the security of the Republic is of infinite analytical weight—the power to wage war is the power to wage war successfully—and the lawyers who provide the analysis are in no position to gainsay policy makers who insist that a given policy, a given drone strike, is necessary to prevent a threat to the Republic. Constitutional law, unlike (say) international humanitarian law with its requirement of proportionality, provides no conceptual tools for dealing adequately with question whether U.S. policy or practice is too savage to be wise or acceptable.

Hughes's opinion for the Court in *West Coast Hotel v. Parrish*, 300 U.S. 379, 391 (1937), struck a similar note in rejecting the due process argument there: "[T]he liberty safeguarded is liberty in a social organization which requires the protection of law against the evils which menace the health, safety, morals, and welfare of the people. Liberty under the Constitution is thus necessarily subject to the restraints of due process, and regulation which is reasonable in relation to its subject and is adopted in the interests of the community is due process." As my citation to *West Coast Hotel* is meant to suggest, however, I think this line of reasoning is properly seen as an argument about *substantive* due process (on which, see the next chapter) rather than about *procedural* due process, which is the concern of the White Paper and the OLC Mem, and mine in this section of this chapter. Treated as a claim about procedural due process I think the argument Professors Bobbitt and Wuerth adopt falls into the error against which Justice O'Connor warned in *Hamdi*, of eroding the "essential constitutional promises" embodied in procedural due process. Congress's enactment of the AUMF affords no individual notice that he is a particular target, nor an opportunity to rebut the government's reasons for contemplating a strike, nor a neutral decision maker to decide if the government is mistaken. I think it better simply to say flat out that we don't pretend to provide procedural due process to an individual targeted for killing in wartime circumstances than to dilute the meaning of the concept.

As we saw above, the nineteenth-century Supreme Court adopted the sensible conclusion that constitutional prohibitions such as the Fifth Amendment Due Process Clause do not constrain the federal government when it directs its war powers against those who may lawfully be treated as enemies.[66] If on the other hand we insist that the usual constitutional prohibitions *must* limit the executive in such circumstances, then it is almost inevitable that the lawyers will redefine the "limits" so they permit what the policy makers think necessary. I do not doubt the sincerity either of President Obama or his legal advisors. The Pickwickian treatment of procedural due process in the Justice Department documents on the killing of Anwar al-Awlaki is the price officials from the president on down have paid (unwittingly, no doubt) for giving themselves, and the rest of us, the reassuring sense that due process is being observed, that what is being done is orderly, civilized, appropriate. But that sense

66. This reasoning applies equally to the Fourth Amendment, which therefore was irrelevant to the decision to target al-Awlaki. If we reject this argument, the Fourth Amendment's prohibition on "unreasonable seizures," which applies to the use of deadly force in the exercise of domestic powers of governance, might apply to the killing of al-Awlaki. *See* Tennessee v. Garner, 471 U.S. 1, 7 (1985) ("apprehension by the use of deadly force is a seizure subject to the reasonableness requirement of the Fourth Amendment"). The OLC Memo assumes the amendment's applicability but correctly observes that the "constitutionality of a seizure is determined by 'balanc[ing] the nature and quality of the intrusion on the individual's Fourth Amendment interests against the importance of the governmental interests alleged to justify the intrusion.'" OLC Memo, *supra* note 7, at 41 (quoting *Garner*, 471 U.S. at 8). The memo reasons that "at least where high-level government officials have determined that a capture operation overseas is infeasible and that the targeted person is part of a dangerous enemy force and is engaged in activities that pose a continued and imminent threat to U.S. persons or interests [REDACTED] the use of lethal force would not violate the Fourth Amendment." *Id.* This conclusion seems correct in light of the Supreme Court's view that in the context of domestic law enforcement, "[w]here the officer has probable cause to believe that the suspect poses a threat of serious physical harm, either to the officer or to others, it is not constitutionally unreasonable to prevent escape by using deadly force." *Id.* (quoting *Garner*, 471 U.S. at 11).

is illusory: Constitutional law, and those who argue over it, can provide no such guarantees in this context, not because in the midst of arms the laws are silent but because constitutional law cannot banish brutality from war. Failing to recognize this basic truth, the Obama administration turned to an antiseptic language about a "balancing of interests" that lacks even the basic candor without which law becomes a meaningless charade: The administration was not trying to respect or take into account al-Awlaki's interests but to destroy them. The killing was constitutional, and it may have been wise, but we should not pretend that it was something it was not.

Chapter 5

Targeted Killing and the Future: Three Speculations

The U.S. government acted consistently with the Constitution of the United States when its executive branch killed Anwar al-Awlaki. That conclusion is limited by the circumstances, factual and legal, of the killing. The war on terror was in what I have called its Phase Three, a time of congressionally authorized hostilities against a specific (if broadly defined) set of enemies. U.S. military and intelligence officials had good reason to view al-Awlaki as the equivalent, in this conflict, to Admiral Yamamoto in the Second World War, a high-ranking enemy combatant and a lawful target. In authorizing the executive to use military force against al-Qaeda and its allies, Congress had placed no geographic limitations on U.S. operations, nor had it attempted to regulate or prohibit targeted killing by drone strike.

All of the factors underlying the conclusion that the al-Awlaki killing was constitutional are changing, even as the executive's reliance on drones seems to be increasing. Phase Three seems to be drawing to a close, giving way to Phase Four, an era in which the war on terror continues but against enemies that did not exist, or had no part in, the attacks of 9/11. In this chapter, we shall examine several constitutional issues that future circumstances may present.

A. WHAT HAPPENS AFTER THE AUMF?

Justice O'Connor's opinion in the detention case *Hamdi v. Rumsfeld* suggests the possibility that the 2001 Authorization for the Use of Military Force (AUMF) limits the hostilities it authorizes in an implicit fashion—temporally.[1] Hamdi argued that even if (erroneously, in his view) the executive were thought authorized to detain him in accordance with traditional law of armed conflict principles, those principles do not allow the indefinite or endless detention of enemy combatants. O'Connor agreed with Hamdi's legal premise: "It is a clearly established principle of the law of war that detention may last no longer than active hostilities." This principle allows "indefinite" detention, but only in the sense that "the date on which the conflict will end" is uncertain: The law of armed conflict assumes, however, that any conflict will come to an end and thus that the authority to detain will eventually terminate. However, as Hamdi pointed out and Justice O'Connor agreed, the current "war on terror" may not fit within these traditional assumptions because it may never come to a formal end. "[W]e understand Congress' grant of authority for the use of 'necessary and appropriate force' to include the authority to detain for the duration of the relevant conflict, and our understanding is based on longstanding law-of-war principles"—and thus that the AUMF presupposes that the authority it confers will eventually come to an end. "If the practical circumstances of a given conflict are entirely unlike those of

1. I have borrowed this title, and learned a great deal, from an excellent article by Jennifer Daskal and Stephen I. Vladeck. *See* Jennifer Daskal & Stephen Vladeck, *After the AUMF*, 5 HARV. NAT. SEC. J. 115 (2014).

the conflicts that informed the development of the law of war, that understanding may unravel."[2]

The current conflict with al-Qaeda and its allies clearly does present the possibility that what Justice O'Connor called "long-standing law-of-war principles" are going to prove inadequate to resolve the question of how long the United States may detain those whom it has captured. Statements by current officials, including the president, give mixed signals about the extent to which the Obama administration believes that the conflict is winding down,[3] and despite the president's clear desire to find a termination point for the AUMF and the state of limited war that it created, it also seems to be the case, unsurprisingly, that executive branch policymakers believe it will be necessary for the foreseeable future to use military means from time to time to address particular terrorism threats. In 2013, for example, a senior Defense Department official testified before a congressional committee last year that "U.S. military operations against al-Qaida and associated forces [will remain necessary] 'for quite a while. . . beyond the second term of the president. . . . I think it's at least 10 to 20 years.' "[4] According to that official, "the AUMF as it is currently structured" provides adequate legal authority for this extended period of military action—perhaps extending

2. 542 U.S. 507, 520–21 (2004) (plurality opinion).

3. *See, e.g.*, Remarks by the President at a Press Conference (Aug. 9, 2013), https://www.whitehouse.gov/the-press-office/2013/08/09/remarks-president-press-conference: "So it's entirely consistent to say that this tightly organized and relatively centralized al Qaeda that attacked us on 9/11 has been broken apart and is very weak and does not have a lot of operational capacity, and to say we still have these regional organizations like AQAP that can pose a threat, that can drive potentially a truck bomb into an embassy wall and can kill some people."

4. *See* http://www.salon.com/2013/05/17/pentagon_official_drone_policy_should_remain_for_at_least_20_years/ (quoting testimony by Assistant Secretary of Defense Michael A. Sheehan before the Senate Armed Services Committee).

beyond a third of a century after the passage of the AUMF.[5] But is this view of the AUMF's indefinite temporal extension persuasive?

In the circumstances of *Hamdi* itself, Justice O'Connor found it unnecessary to solve the legal conundrum potentially created by an endless war waged by authority traditionally understood to come to an end: "Active combat operations against Taliban fighters apparently are ongoing in Afghanistan. . . . If the record establishes that United States troops are still involved in active combat in Afghanistan, those detentions are part of the exercise of 'necessary and appropriate force,' and therefore are authorized by the AUMF."[6] Generalizing from the specific issue before the *Hamdi* Court, the interpretive problem is whether the AUMF can properly be construed to authorize the executive to use military force in perpetuity, so long as no president announces victory. As of the time of writing, the active U.S. combat role on the ground in Afghanistan has ended, the Obama administration apparently believes that the United States is close to the point where the original al-Qaeda network can be thought destroyed, and much of the actual focus of U.S. military action has shifted to the Islamic State and other post-9/11 threats.[7] These changes will soon make it impossible to avoid addressing the question *Hamdi* avoided: Can the executive rely in good faith on the 2001 AUMF as legislative authorization for the use of military force today and in the future?

There is a straightforward legal argument that the AUMF remains in effect until Congress amends it or the president determines that its purpose ("to prevent any future acts of

5. *Id.*

6. 542 U.S. at 521 (plurality opinion).

7. *See, e.g.*, Statement by the President on ISIL (Sept. 10, 2014), https://www.whitehouse.gov/the-press-office/2014/09/10/statement-president-isil-1.

international terrorism against the United States by" the enemies involved in 9/11) has been achieved. The AUMF has no express sunset provision, and it includes no requirement that the president periodically certify that its authorization for the use of force remains necessary or appropriate as is required by some emergency legislation.[8] The objective that Congress directed the president to pursue invites a judgment about the world: Have we in fact so damaged or destroyed the statutory enemies that they lack the ability to carry out "future acts of international terrorism against the United States?" Unless and until the president determines that this objective has been met, it is almost impossible to imagine a court entertaining the issue, and while Congress leaves the AUMF in place one might infer that it does not dispute the executive's judgment.[9]

This is a powerful line of reasoning, and it might be correct, but I do not think that it is the best reading of the AUMF. Consider the following. Congress's power to authorize and limit hostilities is substantive, as we have seen, and not merely the authority to declare a formal state of war; furthermore, Congress's

8. *See* David Abramowitz, *The President, the Congress, and Use of Force: Legal and Political Considerations in Authorizing Use of Force Against International Terrorism*, 43 Harv. Int'l L.J. 71, 75–77 (2002) (relating the discussions in Congress that led to the decision not to include such requirements in the AUMF).

9. Adrian Vermeule's powerful article on Justice Holmes's view of the limits of emergency power demonstrates that it is not in fact completely impossible to imagine the existence of a justiciable temporal limitation on statutory authority that is predicated on the existence of exigent circumstances. *See* Adrian Vermeule, *Holmes on Emergencies*, 61 Stan. L. Rev. 163 (2008). However, Professor Vermeule also notes wryly that "[p]erhaps the Holmesian approach works only in the hands of a Holmes." *Id.* at 200. I do not know whether this Professor Vermeule is acquainted with the Professor Vermeule who co-authored a book extolling the virtues of a legally-unbound executive. *Compare* Vermeule, *id.* at 189 (for Holmes, whom Vermeule views as "our greatest judge," the "guiding assumption may have been that even during war . . . statutes and treaties have the final word") *with* Eric A. Posner & Adrian Vermeule, The Executive Unbound: After the Madisonian Republic (2010), at any randomly chosen page. *See also* the Appendix to this book.

substantive power takes precedence over the president's unilateral power as commander in chief. The asymmetrical constitutional relationship between the superior (although not exclusive) legislative power to determine the scope of U.S. military action and the president's power to direct military action within the scope authorized is thus a key feature of the Constitution's separation of powers. Adherence to these principles in drawing conclusions about the exercise of the federal government's war powers is crucial because the president's invocation of such powers puts at risk "the balanced power structure of our Republic . . . the equilibrium established by our constitutional system."[10] It is well-established that the political branches are not at liberty to vary such constitutional structures even if they agree on the value of the change,[11] and the "avoidance canon" would be appropriately invoked to construe an ambiguous statute not to erode significantly the Constitution's dispersal of authority over the scope, including the temporal scope, of hostilities.[12] The AUMF is ambiguous in this regard: By defining a specific objective for the president's use of force, it implicitly suggests that at some point the enemies involved in 9/11 (including those sheltering the original participants) will be defeated or simply no longer exist. Construing the AUMF as permanently authorizing the use of force against international terrorism generally in the absence of a formal surrender by the specific enemies the AUMF designates is arguably inconsistent with its purpose language as well as with what Congress must be assumed to have understand about the nature of at least some of the enemies it targeted (i.e., that they

10. 343 U.S. at 634, 638 (Jackson, J., concurring).

11. *See, e.g.*, Clinton v. City of New York, 524 U.S. 417 (1998) (invalidating the Line Item Veto Act).

12. On the avoidance canon of statutory construction, *see, e.g.*, Stern v. Marshall, 131 S. Ct. 2594, 2605 (2011) (the Supreme Court "will, where possible, construe federal statutes so as 'to avoid serious doubt of their constitutionality' ").

were highly unlike to surrender or agree to a cease-fire in the traditional sense). Since Congress wisely authorized hostilities against specific enemies rather than war on an idea,[13] perhaps it is appropriate to construe the AUMF to lapse when its factual predicate, the existence of a terrorism threat posed by those specific enemies, has ceased to exist.

The rejoinder is obvious: Even if it would be wrong to construe the AUMF as *formally* authorizing hostilities in perpetuity, only the president (or Congress, should it choose to override the president) can determine when *in fact* the AUMF's objective has been achieved, and it will never be possible for any president to make that determination with perfect certainty. The nature of international terrorism in the twenty-first century is such that no sharp lines can be drawn between causes, organizations, and individual terrorist activities. There is no reason to doubt that as a matter of fact there will be historical links between the original al-Qaeda and its actions in 2001 and the activities of a terrorist organization in 2051 that has no actual interest in al-Qaeda, its causes and beliefs, or the events of 9/11, and yet poses a threat to the United States. If the existence of such factual or historical links is sufficient to bring the latter within the scope of the AUMF, then in practice (given the realities of twenty-first-century terrorism) Congress has authorized permanent war. But as President Obama has affirmed, the state of war has not been our constitutional norm. "I look forward to engaging Congress and the American people in efforts to refine, and ultimately repeal, the AUMF's mandate. And I will not sign laws designed to expand this mandate

13. See Grenville Byford, *The Wrong War*, FOREIGN AFF. 34 (July/Aug. 2002) ("Wars have typically been fought against proper nouns (Germany, say) for the good reason that proper nouns can surrender and promise not to do it again. Wars against common nouns (poverty, crime, drugs) have been less successful. Such opponents never give up. The war on terrorism, unfortunately, falls into the second category.").

further. Our systematic effort to dismantle terrorist organizations must continue. But this war, like all wars, must end."[14]

The legal difficulty with drawing an implicit temporal limitation to the AUMF's authorization of hostilities is parallel to the difficulty in tort law that is addressed by distinguishing cause-in-fact and proximate cause: In order to maintain a legal principle of importance (tort liability should not be so expansive as to render all activity perilous), we cut off a potentially endless expansion of liability by imposing a legal limitation on the potentially endless factual chain of causation. A somewhat similar limitation, imposed by the executive itself rather than by a court, would be appropriate with respect to the AUMF. At some point, it will be possible to recognize that the connections between the current organizations engaged in international terrorism, and the actors and events of 9/11, have simply become too attenuated to sustain a good faith claim that the use of force against one of those organizations genuinely addresses the original statutory purpose of eliminating the threat of terrorism by *only* "those nations, organizations, or persons he determines planned, authorized, committed, or aided the terrorist attacks that occurred on September 11, 2001, or harbored such organizations or persons." When that point is reached, the definitional limitation imposed by the AUMF on what enemies may be fought ought to come into play: Those enemies would no longer exist as a practical matter even if other terrorist organizations with historical connections to the al-Qaeda of 9/11 do. My suggestion here is that the goal of preserving "the balanced power structure of our Republic" but also the interests of a conscientious president seeking to understand the

14. Http://www.aclu.org/files/assets/obama-speech.pdf . On the cultural and constitutional significance of the distinction between war and peace as temporal categories, *see* Mary L. Dudziak's remarkable book *War•Time: An Idea, Its History, Its Consequences* (2012).

scope of his or her authority, would be served by acknowledging a temporal limitation to the AUMF's authorization of hostilities that is implicit in the statute's designation of the enemies against whom military force may be applied.

We may already be approaching the point at which the executive branch ought to acknowledge a duty to recognize this sort of temporal limitation on the scope of authority that the AUMF conveys. As the strategic circumstances change in fact, in part because of the executive's success in executing the AUMF, to a detached observer, the contemporary threat of international terrorist attacks against the United States or its interests and allies will appear increasingly unrelated to the plans and deeds of Osama bin-Laden and his associates.[15] Within the executive branch of the U.S. government, on the other hand, it may always be possible to discern, in good faith, a connection between the enemies of 9/11 and the enemies of today.[16] The president and his advisors, the armed forces and the intelligence agencies, after all, are and ought to be vigilant and proactive in discerning and attempting to counter threats to the security of the nation: As Justice Souter observed in *Hamdi v. Rumsfeld*, it is "the Executive Branch of Government, whose particular responsibility is to maintain security." But the very fact that the executive is "the

15. *See, e.g.*, http://www.volokh.com/2013/03/04/us-surveillance-drones-aid-french-airstrikes-in-mali/ (visited May 17, 2013) (discussing U.S. surveillance-drone support for French operations in Mali and the possibility that at some point "those surveillance drones [will] turn into attack drones").

16. The use of the name "al-Qaeda" is a good example. It is difficult to fault policymakers who see a terrorist organization that applies that label to itself as an enemy within the scope of the AUMF. In fact, however, there are no rules of trademark infringement in the world of international terrorism, and something calling itself "al-Qaeda" may have no significant factual link to the "virtual state" over which bin-Laden once presided. *See* Dana Milbank, With al-Qaeda, What's in a Name? (Jan. 14, 2014), https://www.washingtonpost.com/opinions/dana-milbank-with-al-qaeda-whats-in-a-name/2014/01/14/15dfa354-7d5b-11e3-93c1-0e888170b723_story.html .

branch of the Government asked to counter a serious threat" means that "[f]or reasons of inescapable human nature," executive officials will be tempted to stretch any legal authority they believe the president possesses to meet "the claim[s] that security legitimately raises."[17] The Obama administration's efforts to shoehorn military action against IS into the legal authority conveyed by the AUMF, discussed in Chapter One, illustrates the phenomenon, and as I pointed out helps to create a murky Phase Four in which it is as difficult to describe the president's constitutional authority as it is to determine the factual connections among various organizations engaged in terrorism. Acceptance by the executive branch of a temporal limitation on the AUMF (something like "until we can no longer say that our current concerns are related to the enemies of 9/11 except historically") would serve the principle that on "any reasonable view of separation of powers" the law should not be construed in a fashion that "serves only to *condense* power into a single branch of government."[18]

If at some point the executive branch must recognize that the AUMF no longer provides statutory authority for operations against international terrorists, what will be the constitutional state of affairs? Will we have returned to the situation in what Chapter One termed Phase One, with the president free of any specific legislative constraints on his freedom of action other than those implicit in the limited military means Congress is willing to create and fund? Presumably that would be the correct legal analysis if Congress were to repeal the AUMF without providing any substitute legislation. A simple repeal would withdraw the executive's statutory authority, but under existing Supreme Court

17. *Hamdi*, 542 U.S. at 545 (Souter, J., concurring in part, dissenting in part, and concurring in the judgment).

18. I borrow language from Justice O'Connor, writing in a not wholly-dissimilar context. *See Hamdi*, 542 U.S. at 536 (plurality opinion).

case law it would do nothing more: The congressional silence as to postrepeal presidential decisions would be just that, a failure on the part of the legislative branch to act rather than a restriction on his or her discretion.[19] The termination of authority under the AUMF by expiration, however, might point to a different conclusion. As shown in Chapter Four, the AUMF limits presidential authority to take military action against international terrorism by defining what his authority is. Within the general scope of the statute, therefore, the president is bound by its terms and cannot invoke whatever authority Article II of the Constitution might afford him on its own. It could be argued that construing the AUMF to include an implicit temporal limitation—you may use military force against the statutory enemies until you have achieved the statutory purpose of preventing further attacks by those enemies—implies a prohibition on similar military action after expiration in the absence of further legislation.[20]

19. In the Steel Seizure Case, several of the Justices in the majority, including Justice Jackson, read Congress's failure to provide statutory authority for the preemptive seizure of a vital industry threatened by a shutdown as an implicit prohibition on President Truman's action. As Justice Frankfurter put it, "Congress has expressed its will to withhold this power from the President as though it had said so in so many words." Youngstown Sheet & Tube Co. v. Sawyer, 343 U.S. 579, 602 (1952) (Frankfurter, J., concurring). To the extent that these Justices may have thought Congress's mere silence could amount to a statutory ban on action that the president might otherwise take, their view would be obsolete under more recent case law holding that Congress's only lawful means of adopting rules with the force of law is through bicameral passage and presentment to the president of ordinary legislation. *See, e.g.*, INS v. Chadha, 462 U.S. 919, 944 (1983). In the context of *Youngstown* itself, their conclusion remains defensible and indeed (I think) persuasive. Congress had enacted elaborate, cumbersome, and time-consuming procedures to address the sort of situation Truman faced. It is an eminently sensible construction of those statutory provisions to read them as implicitly forbidding a presidential end run around the prescribed procedures.

20. The reader may recall from Chapter One that an actual attack on the United States itself would give rise to the unquestioned and very broad authority to respond to the attack that President Bush possessed during the brief Phase Two of the present conflict. It would be absurd to construe a statute to limit such authority in the absence of an unmistakable congressional intent to do so.

There is a certain dry logic to this argument, but the logic belongs to an academic exercise, not that to be expected in practical instruments of governance. The AUMF ought to be construed to expire at some point even though it has no sunset provision because the practical consequences of not giving it that reading risk serious damage to the constitutional distribution of war powers between Congress and the president. The practical consequences of construing the AUMF's expiration to forbid executive action against organizations akin in nature or purpose to al-Qaeda would also be serious, but for the opposite reason: Without making a deliberate and express decision to do so, Congress would have disabled the president from making unilateral use of military force against such potential threats even if Article II of the Constitution would otherwise have allowed the president to do so. Quite apart from the implausibility of that interpretation of congressional will, giving the expiration of the AUMF such a construction would tilt the default constitutional balance of power and responsibility between the political branches. Within broad limits, Congress can tilt the balance, as we saw in Chapter Three, but "if the contours of the divided war power contemplated by the framers of the Constitution are to remain," legislation should be construed to respect the Constitution's recognition of "the need for quick executive response to rapidly developing international situations."[21] To the uncertain extent that Article II authorizes military action by unilateral presidential decision, after the expiration of the AUMF, the president will once again possess that authority with respect to international terrorists. We will have returned, constitutionally, to Phase One.

21. *The President and the War Power: South Vietnam and the Cambodian Sanctuaries*, 1 Op. O.L.C. Supp. 321, 331–32, 321 (1970).

B. CAN CONGRESS BAN TARGETED KILLING BY DRONE?

It is widely though not universally believed that congressional interference with the tactical decisions of the commander in chief and his or her uniformed subordinates is an unconstitutional intrusion into the exclusive command functions of the president.[22] This position has to be reconciled with the principle that Congress need not give the president a military blank check but may limit the scope of the hostilities it permits: In the Quasi-War with France that gave rise to *Bas*, *Talbot*, and *Little*, for example, Congress limited the president to the use of naval power, and indeed to only certain naval actions. The difficulty of specifying in advance, or even at the time of decision, where to draw the line between permissible legislative limitations on the hostilities authorized, and impermissible legislative micromanagement of the commander in chief, does not imply that there is no line. Like many other issues in law, the difference is one of degree, not of contrary, hermetically sealed categories.[23] In what follows, I assume that the president has a "conclusive and exclusive ... command function" and that Congress may impose binding limitations on the president's exercise of the command function. Would it be constitutional for Congress to enact legislation

22. *See, e.g., The President and the War Power: South Vietnam and the Cambodian Sanctuaries*, 1 Op. O.L.C. Supp. 321, 334 (1970) (noting the "exacerbated" constitutional problems with a congressional "attempt by detailed instructions as to the use of American forces already in the field").

23. *Cf.* Missouri, Kansas, & Texas R. Co. v. May, 194 U.S. 267, 269 (1904) (Holmes, J.) ("The question is whether this case lies on one side or the other of a line which has to be worked out between cases differing only in degree."); *Civil-Service Commission*, 13 Op. Att'y Gen. 516, 525 (1871) ("the difficulty of drawing a line [between categories] is no proof that both classes do not exist. In constitutional and legal inquiries, right or wrong is often a question of degree.").

limiting or even prohibiting targeted killing whether pursuant to a statutory authorization for the use of military force or as a regulation of the president's exercise of unilateral Article II powers?

Let us put to one side a situation in which Congress forbade the purchase of weapons-capable drones. At some point of equipment failure, this would have the practical effect of banning drone strikes because only Congress can raise and authorize the expenditure of federal funds. However, as in many other situations in which the power of the purse formally provides Congress with the means of exercising control over the executive, in practice this is probably an almost entirely theoretical possibility: What congressional majority would ever be willing to deprive the U.S. armed forces of a cutting-edge weapons system that reduces the extent to which military personnel are actually exposed to harm? A ban on the expenditure of appropriated funds to carry out drone strikes would be effective only if the executive branch accepted the constitutionality of a direct congressional limitation on the use of targeted killing, since otherwise the executive's lawyers would almost certainly view a limitation couched in terms of spending as an unconstitutional condition on the president's discretion over military spending and advise the president that he or she could disregard it. The realistic constitutional question, therefore is the extent to which Congress has the substantive authority to override the commander in chief's discretion as to the use of targeted killing as a means of carrying on hostilities that are themselves lawfully authorized by statute or Article II.[24]

24. Since, as I noted above, to be lawful under international humanitarian law *any* military killing must be targeted in a very real sense, there might be serious difficulties with drafting legislation that accomplished precisely what Congress intended. For the purposes of this discussion, let us imagine a fairly straightforward example: a prohibition on the use of munitions fired by unmanned aerial vehicles (drones) for the purpose of killing human beings even if those persons would otherwise be lawful targets of the use of lethal force.

It is clear that Congress has the power to forbid conduct by the U.S. armed forces that involves the sort of criminal behavior traditionally forbidden by such sources of domestic law as the Uniform Code of Military Justice and the Torture Act.[25] This is not because murder, rape, torture, and the wanton destruction of private property cannot be tools of military force: They have often been used as such, quite deliberately and at times effectively. They are objectionable not because they cannot be rational means to the achievement of military objectives, but because they are wrong.[26] U.S. law has forbidden such *malum in se* behavior by military personnel since the beginning of the Republic, and authority to legislate for that purpose fits comfortably within the delegation to Congress of power to "make Rules for the Government and Regulation of the land and naval Forces."[27] It is settled law that the president has no constitutional authority to order members of the armed forces to violate legislation valid under the Military Government Clause.[28] There are serious arguments that targeted killing is immoral even within the harsh world of military conflict, and indeed that international humanitarian law ought to ban its employment, just as the use

25. The infamous 2002 Office of Legal Counsel (OLC) interrogation memorandum that gave an implausibly narrow construction to the Torture Act based its "reading" of the act on the ground that to give the act a broader interpretation would be to infringe on an area of exclusive presidential authority, but its constitutional analysis was radically defective. *See* H. JEFFERSON POWELL, THE PRESIDENT AS COMMANDER IN CHIEF: AN ESSAY IN CONSTITUTIONAL VISION 27–52 (2014).

26. *See generally* CHARLES FRIED & GREGORY FRIED, BECAUSE IT IS WRONG (2010). In the sentence in the text I am using the term "rational" in the narrow sense of "free from logical or pragmatic self-contradiction," not as a synonym for "reasonable" in the sense of conforming to some normative standard of what makes good sense.

27. U.S. CONST. art. I, § 8, cl. 14.

28. *See, e.g.*, Loving v. United States, 517 U.S. 748, 767 (1996) ("Under Clause 14, Congress, like Parliament, exercises a power of precedence over . . . Executive authority.").

of chemical weapons has long been banned by treaty and is now part of customary international law.[29] If a congressional majority came to that moral conclusion and successfully enacted a ban on military personnel (and all other federal employees, to avoid a CIA loophole) carrying out targeted killings, would the ban be constitutional and binding on the president as commander in chief?

The argument from *McCulloch v. Maryland* that such a ban would be valid is straightforward. The constitutional text "does not profess to enumerate the means by which the [legislative] powers it confers may be executed; nor does it prohibit," on its face, a statutory ban on targeted killing, if Congress deems that ban "essential to the beneficial exercise of those powers." The "end" of preventing inhumane and morally intolerable actions by the military forces of the United States is, surely, both "legitimate" and "within the scope of the constitution." The specific question whether a statutory ban on targeted killing is an "appropriate" means, "plainly adapted to that end," could have, I think, only one answer in a court following *McCulloch*. To the argument that the ban was overly broad, unduly restrictive of the president's discretion, or simply unnecessary to avoid the commission of inhumane acts, I think a judge would be obliged to respond that once it is established that the ban is "conducive" to the end of regulating the conduct of military personnel as Congress sees fit, the "degree of its necessity . . . is to be discussed in another

29. *See* Bond v. United States, 134 S. Ct. 2077, 2083-84 (2014) (discussing history of treaties banning chemical weapons); International Committee of the Red Cross, Customary International Humanitarian Law, commentary on Rule 74, https://www.icrc.org/customary-ihl/eng/docs/v1_cha_chapter24_rule74 (noting that "State practice establishes this rule [banning chemical weapons] as a norm of customary international law").

place," in other words by Congress, and perhaps by the president in exercising the power to sign or veto the legislation.[30]

The argument against the validity of our imagined ban is also straightforward. On the assumption that the "command function" of the commander in chief is exclusive and preclusive of congressional interference, a modest definition of that exclusive presidential power would view it as "measured by the command functions usual to the topmost officer of the army and navy."[31] Rather than being tautological, such a definition has substance when one thinks about it from the standpoint of constitutional purpose (*McCulloch*!). The president's "command power," on the one hand, "is not such an absolute as might be implied from that office in a militaristic system but is subject to limitations consistent with a constitutional Republic whose law and policy-making branch is a representative Congress." But equally, it is the president whom the Constitution authorizes to exercise the "exclusive function to command the instruments of national force. . . when turned against the outside world for the security of our society." The president cannot rightly invoke the Commander in Chief Clause as justification for a military policy of murder or torture for the same reason he or she cannot unilaterally decide to start a war of conquest—those are not the sort of decisions that "the topmost officer" of a branch of the military can legitimately make in our kind of constitutional and civilized Republic.

The decision to kill lawfully targetable enemy combatants through one means rather than another is just the sort of decision generals and admirals must make in times of war, and there is no legal or moral reason why they cannot decide that it

30. McCulloch v. Maryland, 17 U.S. (4 Wheat.) 316, 408, 417, 421, 415, 423 (1819).

31. *Youngstown*, 343 U.S. at 645 (Jackson, J., concurring).

is appropriate to kill a specific enemy combatant, for example because of his or her crucial role in the enemy command structure. Drone strikes are not subject to a per se ban in international humanitarian law, and there seems to be no theoretical reason why they necessarily violate the targeting principles recognized there and treated as obligatory by the U.S. armed forces. In many situations, the use of a drone strike is in fact less likely to lead to unintended harms (in the ugly jargon, "collateral damage") and will minimize the risks to U.S. personnel.[32] An attempt by Congress to forbid the president from making this kind of command decision, at least in the absence of an international consensus that targeted killing by drones is a moral evil similar to the use of chemical weapons and therefore illegal per se, is an unconstitutional attempt by Congress to usurp the president's exclusive role.[33]

In my judgment these arguments are so nearly in equipoise that I would have no confidence in choosing between them except for a background principle which I readily accept: Any presidential discretion to decline to execute an act of Congress that isn't patently invalid under Supreme Court precedent must be as tightly constrained as is possible consistent with respecting the indubitably exclusive and preclusive presidential power that Congress is supposedly infringing. I accept the existence of an exclusive commander in chief command function, but I am not persuaded that conceding Congress's power under the Military Government Clause to ban targeted killings unduly compromises

32. *See, e.g.*, Daniel Byman, Why Drones Work (July/August 2013), https://www.foreignaffairs.com/articles/somalia/2013-06-11/why-drones-work.

33. It is unnecessary, I think, to prolong this discussion by considering the constitutionality of legislation that regulated targeted killing in specific ways without banning it altogether, for example by prohibiting the targeted killing of U.S. citizens who would otherwise be lawfully subject to lethal military force. The basic arguments for and against the validity of such regulations would be the same.

that exclusive presidential function, and I do think that denying Congress's power on the matter gives the Military Government Clause a more parsimonious reading than seems entirely consistent with *McCulloch v. Maryland*.

We can consider more briefly a second argument for congressional authority to ban targeted killings. As we have seen, Congress has the authority not just to authorize hostilities but also to limit them. As Justice Chase wrote long ago in *Bas v. Tingy*, Congress may authorize acts of war that are "limited in place, in objects, and in time," and "if a partial war is waged, its extent and operation depend on our municipal laws."[34] Would it be constitutional for Congress to ban the use of targeted killing by defining the hostilities it is authorizing so as not to include drone strikes, just as the Congress of 1798 banned the use of the U.S. Army to invade French colonies by defining the use of armed force it was authorizing to "hostilities on the high seas by certain persons in certain cases"? The answer found more plausible is, I think, going to depend almost entirely on how broadly one understands the president's exclusive command function, which was the determining factor, for me at any rate, in thinking about a prohibition based on the Military Government Clause. Forbidding the president to use this particular weapons system, in this particular manner, seems *to me* more akin to the sort of limitation on hostilities that the old Supreme Court decisions accepted as binding on the executive, than it does to legislation dictating which among the existing range of infantry weapons the Marines are to use on the battlefield, which I think would infringe the exclusive command function.

In 1798 Congress wanted to deter French aggression with a measured degree of force that would not be likely to escalate or to

34. Bas v. Tingy, 4 U.S. 37, 43 (1800) (seriatim opinion).

involve the United States in the wider European conflict; the decision how to pitch the use of force at just the right level was a matter of legislative judgment, and Congress's decision was legally binding on the president regardless of the latter's views about whether Congress's limitations were wise. A present-day Congress might have very similar concerns. The current targeted killing policy, and any likely future executive policy, allows for drone strikes in countries whose sovereignty Congress may want to respect, risks international and political repercussions that Congress may wish to avoid, and might lead to an expansion of the conflict beyond the limits Congress is willing to endorse. Conclusion: The statutory ban would be constitutional. But the arguments that a ban based on the Military Government Clause would be an unconstitutional intrusion on the president's command function are just as cogent here. Readers who accept what I think is the correct tie-breaking principle will conclude, perhaps *dubitante*, that a ban on targeted killings outside the immediate context of a battlefield (where I cannot see a constitutionally cognizable distinction between a drone strike and a sniper's rifle) would be valid; those who don't may very reasonably come to the opposite judgment.

C. ARE THERE CONSTITUTIONAL LIMITS TO THE EMPLOYMENT OF TARGETED KILLING AS A TOOL OF WAR?

The simplest answer is no, not for practical purposes and not when Congress has authorized the president, implicitly or explicitly, to engage in targeted killing. A drone strike in the authorized war might still be illegal under international law, or grossly immoral, or strategically foolish, but all of those can be true and the action in question still lawful under the Constitution.

To be sure, this stark answer needs two qualifications. Of course the powers delegated to Congress and the president are subject to override by applicable constitutional prohibitions—prohibitions trump authorizations—but if the argument in Chapter Three is correct, then the Bill of Rights and other constraints on the federal government's powers of governance simply do not apply to the war powers when the latter are directed to the external world. Even if that argument is incorrect, the common view is that Bill of Rights provisions do not apply abroad, or to the extent they do, they protect only U.S. citizens, which sharply reduces their practical significance: al-Awlaki was a U.S. citizen, but most enemy combatants are not and will not be. The Constitution's prohibitions on the exercise of federal authority are an almost entirely theoretical limitation on the war powers.

But what about the conceptual limitations built into the textual delegation of, say, the power to declare war or to act as commander in chief? It is an unshakable premise of U.S. constitutional law that the Constitution's enumeration of specific powers "presupposes something not enumerated"; the text delegates various specific powers related to war and military matters rather than a general "power to do anything, anywhere, that can be done with an army or navy" or other military means.[35] It would be unconstitutional for Congress to "declare war on crime" and by so doing authorize the president as commander in chief to use drone strikes at his discretion against ordinary domestic criminals: Federal domestic criminal law enforcement is constitutionally authorized only as to matters within Congress's powers of governance and must comply with the Bill

35. Gibbons v. Ogden, 22 U.S. 1, 195 (1824); Youngstown Sheet & Tube Co. v. Sawyer, 343 U.S. 579, 642 (1952) (Jackson, J., concurring).

of Rights.[36] More generally, as Justice Jackson wrote specifically about the "military powers of the Commander-in-Chief," the federal war powers are not meant "to supersede representative government of internal affairs" in accordance with the usual constitutional rules.[37]

This is all true and important . . . and quite irrelevant if the question is whether there are constitutional limits to targeted killing by drone in the war on terror, at least in Phase Three circumstances where Congress has authorized military action, or in a future Phase Two, where the president is responding to a national emergency in advance of legislative approval. In those situations, the federal war powers are at their zenith, and there is no separation of powers issue created by disagreement between the political branches. The argument that in such circumstances there can be no constitutional objection to targeted drone strikes is straightforward and is likely the correct answer to the question posed in this section. A constitutional answer that is "likely" correct is not necessarily so, and there are two speculative constitutional limitations on targeted killing that we shall briefly explore.

Is it possible that the Constitution forbids the use of assassination by the U.S. government? The blogosphere is currently replete with charges that the executive's current practice of targeted killing is nothing more than a series of assassinations, indistinguishable

36. The Supreme Court's Civil War–era decision that the executive branch acted constitutionally in using military force against the Confederacy rested on the analogy between a foreign invasion and the secessionists' military actions against the United States. In either circumstance,

> the President is not only authorized but bound to resist force by force. He does not initiate the war, but is bound to accept the challenge without waiting for any special legislative authority. And whether the hostile party be a foreign invader, or States organized in rebellion, it is none the less a war.

Prize Cases, 67 U.S. 635, 668 (1862).

37. *Youngstown*, 343 U.S. at 644 (concurring opinion).

from killings that we would unhesitatingly give that label and condemn as wrong: Even relatively neutral references to the killing of al-Awlaki often describe it as an assassination.[38] More soberly, some have worried that the policy violates either international law or the executive order providing that "[n]o person employed by or acting on behalf of the United States Government shall engage in, or conspire to engage in, assassination."[39] As to targeted killings undertaken in the current Phase Three, the argument that targeted killings such as that of al-Awlaki are "assassination" as a matter of definition is weak. The scope of the international law prohibition is not entirely clear, but its historical core meaning lies in the ethical and prudential concerns about treacherous behavior that Grotius articulated at the dawn of thinking about the law of war.[40] A drone strike shares the element of surprise with a classical assassination, but there is no obvious sense in which it is a breach of trust. That the strike is targeted at a specific person does not make it a prohibited assassination: "If . . . the target is lawful, for example because the target is directly participating in hostilities or is a member of the armed forces, the fact that he has been targeted individually for attack does not render the attack unlawful."[41]

The executive order banning assassination is similarly uncertain in its range of applications—the order does not define its crucial term—but many careful descriptions of the order's scope limit it to killings for a political rather than a military purpose.[42] The

38. *See, e.g.*, Drone Memo Justifying Assassination of U.S. Citizen Released (June 23, 2014), http://www.usnews.com/news/articles/2014/06/23/drone-memo-assassination-us-citizen-anwar-al-awlaki-released.

39. Executive Order 12333 § 2.11.

40. *See, e.g.*, Thomas C. Wingfield, *Taking Aim at Regime Elites: Assassination, Tyrannicide, and the Clancy Doctrine*, 22 Md. J. Int'l L. & Trade 287, 298–305 (1999) (discussing the historical evolution of international law rule against assassination).

41. William H. Boothby, The Law of Targeting 528 (2012).

42. This is the position taken, for example, in the well-known (but not uncontroversial) legal opinion on the scope of the assassination ban written by W. Hays

Justice Department documents defending the lawfulness of the al-Awlaki killing articulate a clearly military purpose, so even if the executive order is relevant in a Phase Three operation against a member of al-Qaeda, on this reading the order would not apply.

On a more fundamental level, it is very difficult to find a convincing basis for discerning a *constitutional* problem with targeted killings by drone when Congress has approved hostilities, since neither international law nor a self-limiting decision by the executive is in itself a norm of U.S. constitutional stature. A constitutionally authorized military action does not become unlawful, as a matter of domestic U.S. law, because it transgresses an international law norm. President Gerald Ford issued the original executive prohibition at a time constitutionally equivalent to Phase One: Congress had not authorized hostilities and there was no national emergency due to an attack on the United States. Even without the president expressly modifying the order (which the president is perfectly free to do), the 2001 AUMF's authorization to use military force would supersede a unilateral presidential self-restriction to the extent that the AUMF and the executive order overlap, if they do.[43] In any event, there is no apparent reason why a drone strike should differ from a sniper's bullet as a tool of war in a lawful conflict. Either might be seen as dishonorable or cowardly from the perspective of a twenty-first-century Homeric hero, but the Constitution does not limit the political branches to military actions that satisfy some code of

Parks. *See* Memorandum from W. Hays Parks, Special Assistant for Law of War Matters to The Judge Advocate Gen. of the Army: Executive Order 12333 and Assassination (Nov. 2, 1989), https://www.law.upenn.edu/institutes/cerl/conferences/targetedkilling/papers/ParksMemorandum.pdf.

43. Acting as commander in chief, the president could choose not to order or permit military operations that implicated the assassination ban, but that would be an exercise of his or her constitutional discretion rather than compliance with a binding rule, even a self-imposed rule.

warriorly conduct. The federal government's war powers convey the authority to wage war successfully, and it is well within the range of reasonable strategic opinion to think that drone strikes are a valuable tool in prosecuting the war on terror. Whether that is true is not a constitutional question.

But what about drone strikes in the Phase Four that is now beginning? As more and more of the executive's attention is directed toward enemies who fit poorly, if at all, under the AUMF, the president's power to order targeted killings will come to depend more and more clearly on his or her unilateral Article II authority, and the possibility will arise that there is a legal limitation on that authority arising out of the Constitution's division of the war powers between the two political branches. Let me start with a premise that I cannot prove but do not doubt: There is no plausible argument that the Constitution grants the president a James Bond–esque "license to kill." My reference to 007 is not meant to be whimsical. I have no doubt that the U.S. government carried out killings during the Cold War that do not fit easily or perhaps at all within a legal framework. (Whether some or all of those actions could be plausibly viewed as congressionally authorized is beyond the scope of this book.) Any such killings were, however, covert and should not be taken as providing a factual basis for the sort of constitutional argument that open and well-known political practice supports. Supreme Court precedent makes it clear that if the president does possess the lawful authority to order the killing of someone without a statutory mandate, that authority must have a convincing basis in a specific provision in Article II.[44]

44. *See, e.g.*, Medellin v. Texas, 552 U.S. 491, 524 (2008) (as a matter of "first principles . . . the President's authority to act, as with the exercise of any governmental power, 'must stem either from an act of Congress or from the Constitution itself'"). In his canonical Steel Seizure opinion, Justice Jackson rejected as an inadequate basis for constitutional argument nebulous invocations of Article II as a whole, or of the first clause of the article, which provides that "the executive

The Commander in Chief Clause is the most obvious textual source of a unilateral executive power to use lethal force abroad.[45] As Chapter Three has shown, however, the clause is not a plenary grant of authority to use military force, including the pinpoint application of such force by drone strike, whenever the president thinks it expedient to do so. The legal justification for a Phase Four targeted killing by drone of a terrorist must fit within the president's controversial, but this book argues valid, authority to make limited use of military force without statutory authorization: Phase Four is, constitutionally, a period of transition back to the historical norm I have labeled Phase One. A Phase Four drone strike would be justified, the argument goes, as simply one of the forms in which the president can initiate hostilities, assuming that the other grounds for unilateral executive action were met (limitation in anticipated scope, duration, and so on). But there is also a counterargument that at least in some circumstances this limited Article II authority

Power shall be vested" in the president. *See Youngstown*, 343 U.S. at 640–41, 646–47 (concurring opinion). The Supreme Court's recent decision in *Zivotofsky v. Kerry* upheld the executive's claim that the president alone may determine whether a U.S. passport recognizes Jerusalem as part of the State of Israel, but in explaining its reasoning the Court was careful to stress the importance of the text of Article II, specifically the Reception Clause, as well as structural and historical concerns. *See Zivotofsky*, 135 S. Ct. 2076, 2084–85 (2015). *See also* U.S. CONST. art. II, § 3 (providing that the president "shall receive Ambassadors and other public Ministers").

45. In discussing this issue in terms of the Commander in Chief Clause I follow what I think is the more common framing of the question. It is my own view that the president's limited but historically important authority to use military force abroad that does not amount to "war" is best understood as a permissible means of carrying out his or her authority over foreign affairs. *See generally* H. JEFFERSON POWELL, THE PRESIDENT'S AUTHORITY OVER FOREIGN AFFAIRS: AN ESSAY IN CONSTITUTIONAL INTERPRETATION (1992). If I am right, however, nothing in the analysis in the text would change substantively.

The president, it should be noted, may have in some circumstances the authority to use lethal force outside the United States for purposes such as criminal law enforcement, the conduct of intelligence operations, and the fulfillment of treaty obligations. In each of these situations, the executive would not be relying on a unilateral Article II power but on the president's authority as chief executive to carry out the law created by statute or treaty.

would not extend to targeted killing, even though it would justify the deployment of U.S. troops or the conduct of air strikes.

To see how this counterargument runs, we must briefly consider the origin of the claim that the president may sometimes use military force without congressional authorization. Its constitutional basis lay as a historical matter in the early presidents' assertion of the power to use traditional forms of military or naval action to safeguard the foreign policy interests of the United States. From George Washington on, the presidents asserted independent authority to oversee American foreign affairs, and their unilateral and guarded employment of military force was as a tool of foreign policy rather than an attempt to circumvent Congress's authority to decide whether to commit the nation to war or other major hostilities.[46] After war broke out between Revolutionary France and Great Britain, President Washington issued a declaration of American neutrality to avoid U.S. entanglement in the conflict. He and his advisors also determined that if necessary the executive would use military force to prevent French privateers from using American ports and thereby drawing down the wrath of the Royal Navy: The cabinet concluded this limited type of military action was, like the neutrality declaration, within the president's discretion even in the absence of statutory authorization.[47]

46. *See* H. Jefferson Powell, The President's Authority over Foreign Affairs: An Essay in Constitutional Interpretation 34–94 (2002) (making the historical case for this assertion). The use of the Army against Native Americans raises legal issues that have no contemporary parallel: Despite their status as sovereigns for some purposes, the Native American nations were not treated as nation-states in the international law sense and lived in areas over which the United States viewed itself as enjoying de jure sovereignty. The tragic injustice of that history is one more illustration of this book's argument that showing the constitutionality of a military action does not prove that the action is wise or ethical.

47. *See id.* at 51–60 (discussing the Washington administration's military measures).

It is unthinkable, on the other hand, that President Washington and his cabinet would have considered the president authorized to send killers abroad to strike down key figures in the French government, or even officers in the French military and naval forces, to advance even the most pressing foreign policy goals. Even if such a killing could have been accomplished without treachery, and thus might not have been "assassination" in a Grotian sense, it seems overwhelmingly likely that Washington and his contemporaries would have thought it a dishonorable act that the Constitution would not countenance. The eighteenth century in Europe, it has been observed, was remarkably free from assassination in the broad sense of the murder of significant figures.[48] There is no reason to doubt that the eighteenth-century founders thought of the U.S. government's legitimate powers in terms of their cultural expectations about political conduct. Even if Congress can dispense with an assassination limit on military action, we might conclude, it would be wrong to construe Article II to permit the president acting alone to do so, and given the functional similarity between contemporary drone strikes and the older methods of eliminating prominent figures, the limit should be understood to ban targeted killing in the absence of legislation.

This line of reasoning is not completely rootless as an argument about the Constitution's implications. While its details are speculative, the basic idea that constitutional thinking about the war powers ought to take into account the assumptions

48. *See generally* Franklin L. Ford, *Assassination in the Eighteenth Century: The Dog that Did Not Bark in the Night*, 120 PROC. AMER. PHIL. SOC'Y 211 (1976). The general European hiatus in political murders from the late 1600s through 1789 extended to North America as well. "In the American colonies, despite the prevalence of mob violence, despite the ferocity of printed attacks on royal officials, no British officer was the object of a lethal assault." *Id.* at 211.

lying behind the text's delegations of authority is sound. If it would have been unthinkable, on founding era assumptions, that the United States should engage in targeted killing in circumstances entirely removed from the battlefield or even a surprise attack on an enemy camp, that at least suggests that the Constitution's grant of war powers should not be taken to permit such killings. The executive order banning assassination, then, might have constitutional underpinnings: The assassination by U.S. agents of a traditional political figure, no matter how threatening to U.S. interests, would not only be shocking but, conceivably, beyond the president's authority over foreign policy to order regardless of whether it violated any domestic criminal statute. But whatever one might make of this argument as a limit on the scope of the president's power to carry out foreign policy in general, this reasoning is quite unconvincing when applied to targeted killing that has a genuinely *military* rationale.[49]

If the president is ever constitutionally authorized to use military force except pursuant to a statute—and all agree that he or she is so authorized in the event of an attack on the United States—to treat the power (in those circumstances in which it exists) as less than fully competent to its purposes is contrary to the basic framework of constitutional thought that *McCulloch* creates. International terrorism in the twenty-first century has aspects unprecedented in U.S. (and human) history: A small organization with no fixed geographical basis to attack or occupy by traditional military means may nonetheless pose a threat that the executive branch may rightly see as military, both in

49. Of course, a president might invoke a plausible military reason as a pretext for a politically motivated killing. But if the president is willing to act in bad faith and can find subordinates to assist, no legal rule by itself is going to stand in the way.

the organization's destructive potential and in impossibility of countering it successfully through traditional law enforcement tools. The scope of the government's constitutional authorization to use the tools of war to address such a threat is as broad as the threat itself. To the extent that Article II delegates to the president authority to initiate hostilities in Phase Four at all, therefore, the president is authorized to employ targeted killing by drone as one of his or her options.

Assassination, as a description of the killing of al-Awlaki, or of the Obama administration's announced policy governing targeted killing in the war on terror, is a constitutionally meaningless term. The fact that the U.S. government admits that it targets specific enemy combatants for elimination by drone strike taps into a deep uneasiness about certain modes of warfare that, as we saw in the introduction, stretches back to the Iliad. But that uneasiness—or moral repugnance, for those to whom it rises to that level—does not make the practice unconstitutional. Targeting individual enemies has been an accepted part of war making since war began, and the Constitution of the United States is not a pacifist document. For many critics of the administration's use of drones, however, what has gone terribly wrong is not overprecision in dealing out death but what they charge is an egregious failure on the part of the administration to limit drone strikes and their consequences to the enemy combatants the administration purports to be targeting.

The factual allegations that are the basis for this line of criticism fall into two basic categories. The first is that the executive branch uses drones to attack persons not on the basis of known identities but because they satisfy various criteria that military and intelligence experts believe are associated with members of al-Qaeda and other enemy combatants. The criteria used in such "signature strikes" are allegedly too crude to avoid high error rates that lead to many noncombatant deaths.[50] The second is that quite apart from causing civilian casualties, the incidence of drone strikes in parts of Yemen and Pakistan is so high that many civilians are suffering from post-traumatic stress disorder and other stress-induced illnesses, in short, that in the those areas the U.S. war on terror is itself terrorizing the civilian population.[51] The executive branch denies both factual charges, insisting that it is only targeting legitimate targets and that the unintended harm to noncombatants is, while tragic, an unavoidable result of the enemy's own use of civilians as camouflage and well within the margin of error acceptable under international law.[52] What the facts are, I do not know: If the charges are fundamentally accurate, it would appear that the U.S. government is violating the international humanitarian law requirements of distinction and proportionality.

Let us assume for the moment that these factual allegations are true. That the United States is acting in contravention of international law does not necessarily mean that the executive branch is violating the Constitution: The war powers of the federal government are the powers to wage successful any war that is thrust upon

50. *See, e.g.*, Jack Serle, Bureau of Investigative Journalism: Drone War Report, January–June 2015 (July 1, 2015), https://www.thebureauinvestigates.com/2015/07/01/drone-war-report-january-june-2015-controversial-signature-strikes-hit-yemen-and-pakistan/;

David Rohde, What the United States Owes Warren Weinstein (Apr. 28, 2015), http://www.theatlantic.com/international/archive/2015/04/warren-weinstein-drones/391655/.

51. *See, e.g.*, Vivian Salama, Death from Above: How American Drone Strikes Are Devastating Yemen (Apr. 14, 2014), http://www.rollingstone.com/politics/news/death-from-above-how-american-drone-strikes-are-devastating-yemen-20140414; Aliya Robin Deri, "Costless War": American and Pakistani Reactions to the U.S. Drone War, 5 Intersect (2012), http://ojs.stanford.edu/ojs/index.php/intersect/article/download/367/167.

52. *See, e.g.*, https://www.lawfareblog.com/text-john-brennans-speech-drone-strikes-today-wilson-center.

the nation, or that the political branches of the government deem necessary to save the Republic. But the distinction between constitutional and international law does not make the latter irrelevant to constitutional thinking. In *Hamdi v. Rumsfeld*, the Supreme Court construed the 2001 AUMF to authorize the executive to capture and detain enemy combatants because those actions, "by 'universal agreement and practice,' are 'important incident[s] of war.'"[53] It is equally a matter of "universal agreement" if not in fact of universal practice that deliberately terrorizing civilian populations is not a legitimate "incident of war" but in fact a serious violation of international law. *Hamdi* therefore suggests that at least while Phase Three lasts and the AUMF governs the executive's actions, it is unlawful as a matter of domestic law for the executive to disregard the international law limits on harm to noncombatants.

The AUMF may expire, as we have seen, or Congress could modify it to permit the executive to engage in targeted killing by drone even if doing so affects civilian populations as critics claim. Doing so, *ex hypothesi*, would still violate international law, and many of us would regard such actions as morally hideous, but would there be a *constitutional* objection? Defenders of Congress and the president would argue that in the hypothesized circumstances the responsible branches would have made a plausible judgment about what must be done to save the Republic, and that the Constitution demands nothing more. As I wrote above, that harsh answer may be correct, but before we leave the matter there, consider the following. The survival of the United States, perhaps unlike that of a nation defined by ethnicity, is not simply a matter of the continued existence of a polity bearing the familiar name and occupying the historical territory. The "United States" *is* that

53. *Hamdi*, 542 U.S. 507, 519 (2004) (plurality opinion of O'Connor, J.) (citation omitted).

political community constituted (and not just governed) by the fundamental principles embodied in its basic legal document. The Constitution potentially authorizes the federal government to act in contravention of peremptory norms of international law if doing so is necessary to achieve the high constitutional purpose of providing for the common defense. But by the same token, is it not a constitutional purpose of at least as great dignity that the federal government respect the self-evident and universal truths of which the Declaration of Independence speaks?

It comes naturally to Americans to think of those truths in terms of individual rights, as indeed the Declaration of Independence itself does, but their constitutional significance is not limited to interpreting the constitutional prohibitions prescribed by the Bill of Rights. The Constitution grants the political branches the powers necessary to make war in order to preserve the Republic and its principles, not to discard those principles in order to win the war at all costs. If it is or ever becomes true that the U.S. government is itself using terror against civilian populations, then and to that extent the government will have abandoned the substance of American constitutional principles.[54] Does the Constitution actually authorize the government to do so? In discussing the legal and strategic framework of the war on terror, Professor Philip Bobbitt argues that the United States must refrain from using terror itself not simply for moral or prudential reasons, but out of fidelity to its fundamental commitments as a political community.

> These are essentially constitutional issues, but they are not so much matters of civil liberties as they are profound questions of the self-definition societies achieve through their

54. The use of force against Native Americans in the nineteenth century is, I think, a historical and often horrifying example. *See supra* note 46.

> constitutional processes in time of war. They are matters of constitutional legitimacy because they are matters of self-respect.[55]

There are significant difficulties with any effort to develop a satisfactory argument that the Constitution does not authorize the political branches to use drone strikes in a manner that terrorizes civilians. Some of the difficulties stem from the fundamental assumption that undergirded Chapter Two: Constitutional law is a reasoned discipline of legal inquiry. If the war powers do have the sort of outer limit under discussion, it must be possible to state that limit in legally cognizable terms and not just as the set of military actions someone finds morally abhorrent. Doing so would require us, in particular, to develop workable criteria for giving operational legal meaning to the concept of "terror": Any plausible constitutional limit to the scope of the war powers must leave room for the very broad discretion entrusted to Congress and the president in these matters and yet enable reasoned debate over a claim that the branches have exceeded the bounds of that discretion.[56] A workable limit would rule out "only the most egregious" decisions and might draw on the decisions addressing "substantive due process" and executive action, which define "the cognizable level of executive abuse of power as that which shocks the conscience . . . and violates the decencies of civilized conduct."[57]

55. Philip Bobbitt, Terror and Consent: The Wars for the Twenty-First Century 533 (2008).

56. *See id.* at 183: "It will be crucial . . . to develop legal indicia that allow one to determine when a state, including one's own state, threatens to become a state of terror. We must be able to say when an act of violence represents a legitimate effort to preclude terror and when it is itself an act of terrorism."

57. As applied to the actions of executive officers, substantive due process "prevent[s] government officials from abusing [their] power, or employing it as an instrument of oppression." Sacramento v. Lewis, 523 U.S. 833, 846 (1998) (citations omitted).

Even on the assumption that it is possible to devise a workable limit on the war powers of the sort under consideration, there is an even more fundamental difficulty about the limit's legitimacy. The fact that any such limit will be the product of a reasoning process involving the overt construction of legal doctrine rather than the simple identification of preexisting meaning in the relevant constitutional texts should not trouble anyone who recognizes the central role *McCulloch v. Maryland* and the Steel Seizure Case play in constitutional law: It is precisely their fusion of fidelity to constitutional text and purpose with creative thought about the application of constitutional principle that makes the opinions of Chief Justice Marshall and Justice Jackson central to sound constitutional thinking. But in celebrating their creativity, we must not lose sight of their commitment to fidelity. The limit we are considering will only be legitimate as a principle of constitutional law if it is genuinely rooted in that law. It is at this point, I believe, that even someone sympathetic to the argument up to this point may find it impossible to go further. I leave the hesitant—and here I address myself as well as the reader—with a question. The murder and rape of the defenseless have a long history as tools of war, and there is no a priori ground on which to assert that they are never effective: Must we therefore concede that the Constitution of the United States authorizes Congress and the president to employ these tools if they deem it necessary to the defense of the Republic?

Conclusion: Why It Matters

Does any of this matter? Imagine someone deeply opposed to the killing of Anwar al-Awlaki for moral or strategic reasons: For him, a constitutional argument for the lawfulness of the killing might seem worse than useless, a capitulation to something deeply wrong driven by nothing weightier than a notion of "sound" constitutional thinking. Now imagine someone else, just as deeply convinced that the killing of a high al-Qaeda commander was essential to the security of the nation: For her, lawyers' quibbles on the matter might be entirely beside the point in a dangerous world. If custom requires a ritual obeisance to the Constitution, have the lawyers concoct a legal rationale for the killing and let us be done with it.

I am sympathetic to both of my imagined critics, because in part they are both right. The killing of al-Awlaki implicates questions of ethics and the place of the United States in the community of nations that range far beyond the parochial concerns of U.S. constitutional law. Furthermore, lawyers, as lawyers, surely have little enough to contribute when the question is whether a particular military action will enhance American national security: That is a matter for the politicians and the generals. And constitutional law is not Euclidean geometry: At many steps in the argument of the previous four chapters—and certainly as to the central question "was the killing of al-Awlaki constitutional?"—a reasonable person can disagree.[1] It is nonetheless vitally important that we try to get our constitutional thinking right, as opposed to simply arriving at the outcome we think best on some other ground and then contriving a constitutional argument to fit our conclusion. There are at least three reasons why this is so.

The first has to do with the role of law in American society and government. Americans love to hate lawyers and often profess cynicism about the motivations of public servants who claim to act out of constitutional conviction. At the same time American society is notoriously legalistic in its modes of thought, debate, and action. This is nowhere more true than in government, where anxiety over the legality of action is an omnipresent factor. The sources of official worry over legality are complicated, of course, but it is nonetheless true that government officials from the president and members of Congress (and their lawyers) on down must often address concerns, their own or those of others, about what the law is, even if the issue is one unlikely ever to come before a court. The options policy makers consider and those they reject out of hand, the extralegal factors they take into consideration, their final decisions on controversial topics, all reflect the law, or rather what people understand about the law. This can be true at very deep levels when the law in question is the law of the U.S. Constitution.

1. My own view is that, to a much greater extent than is often recognized, there are right answers to constitutional law questions, and that in fact most constitutional problems get resolved in a satisfactory, technical-law fashion in the 99+% of constitutional cases that someone other than the Supreme Court decides. The contrary impression comes from too tight a focus on what the high Court does and from the Justices' own impatience, at times, with the demands of technical reasoning. But nothing in this book rests on whether I am right on this.

A deep concern about the law can serve the political community well in many ways, above all when legislators and executive officers have absorbed the civilizing aspects of the law into their thinking about matters of policy and discretion. But the pervasively legalistic mindset of early twenty-first-century American officials creates a standing temptation to confuse the question of constitutionality with questions of wisdom and morality. The conclusion that X is constitutional slides all too easily into the conclusion that X is a good idea and that qualms or hesitations about X are ungrounded, even though we have learned little relevant to the question whether it would be *good* for the U.S. government to take a particular action—target a U.S. citizen for killing, for example—when we have determined that the action would be *constitutional*. The lawyers can play the vital role of reminding policy makers of the Republic's enduring commitments only if their advice is disciplined by their commitment to sound constitutional thinking. Often enough, as in the case of al-Awlaki, good constitutional reasoning *does* lead to the conclusion that the government can take some action, but trying to get the constitutional analysis *right*, and not just devise a convenient imprimatur for some policy, is a check on what Justice Jackson referred to when he said, "[t]he tendency is strong to emphasize transient results . . . and lose sight of enduring consequences."[2]

The second reason why those who engage in debate over the constitutionality of an act like the killing of al-Awlaki ought to care about getting their constitutional thinking right is that they have a personal duty to do so, at least if they are public officials, or lawyers advising public officials, or (I would add) private individuals acting in their public roles as citizens and voters. Near the

2. Youngstown Sheet & Tube Co. v. Sawyer, 343 U.S. 579, 634 (1952) (concurring opinion).

conclusion of his opinion in *Hamdi v. Rumsfeld*, Justice Antonin Scalia posed and rejected the view Cicero famously articulated that *silent enim leges inter arma* – in the midst of war the laws fall mute. "Whatever the general merits of the view that war silences law or modulates its voice, that view has no place in the interpretation and application of a Constitution designed precisely to confront war and, in a manner that accords with democratic principles, to accommodate it."[3] I pause over the word "democratic," which Americans often invest with a romantic gloss that can mislead, but Justice Scalia was deeply right to insist that in the American constitutional system even war does not set a limit to the reach of the law. No matter how exigent the threat or apparently necessary the action, anyone may question, and a conscientious president must ask, what legal authority the executive possesses to take that action in the defense of the Republic.

The Constitution does not rely entirely on the good faith of the president, or the members of Congress, to be sure: It establishes an elaborate set of institutional arrangements designed to channel and check the exercise of governmental power, and it ultimately assumes that an informed citizenry will exercise its powers to censure and to punish politically wayward governmental actions. But the Constitution presupposes as well that governmental officials can and should act out of a good faith intention to act in accordance with its norms. Constitutionality is a prerequisite of legitimacy in the American political system; disingenuous constitutional reasoning by those charged with making decisions undermines the legitimacy of those decisions and to that extent of the system itself.

As we discussed in the introduction and again in Chapter Two, constitutional analysis is not a matter of algorithmic or strictly

3. *Hamdi*, 542 U.S. 507, 579 (2004) (dissenting opinion).

deductive reasoning. Unlike a formal proof in Euclidean geometry, the formal structure of a constitutional argument does not exclude a role for the exercise of individual judgment: As Chief Justice Marshall wrote long ago, which argument is the most persuasive on a difficult constitutional issue depends, in part, on "the wishes, the affections, and the general theories" of the individual who must decide.[4] Where that sort of judgment must be used, good faith on the part of those exercising judgment becomes essential: Reliance on institutional arrangements alone will not suffice. Judicial review itself, for most Americans the paradigmatic institutional arrangement for checking governmental power, depends on the conscientious exercise of their power by judges, who are themselves, of course, governmental officials.

The governmental system created by the Constitution constantly depends, in fact, on the intelligent and conscientious execution of their duties by those who hold public office and, in the end, those who enjoy the privileges of citizenship. The cynical belief that the Republic can dispense with a starting assumption of good faith is as unrealistic and unworkable as any Pollyannaish call for unguarded trust in the exercise of power. And the form in which those who discuss the Constitution exercise good faith toward the Republic is by making their best efforts to answer constitutional questions with sound constitutional reasoning. Anything else—whether the individual serves the Republic as legislator, executive officer, advisor, judge, litigator, or private citizen—is a cheat.

The third, final reason why sound constitutional thinking matters has to do with the nature of the constitutional enterprise.

4. 4 John Marshall, Life of George Washington 243 (Chelsea House 1983) (1805).

In the introduction I quoted Justice Frankfurter's approving description of the Constitution as "the means of ordering the life of a people," an "instrument of government [rather than] a text for interpretation."[5] The Constitution is surely a text, but it is a text that points beyond itself, serving through the government it creates and limits as the means by which the American political community orders its common life. The Constitution's provisions and principles exist to serve the goals of that common life—which is one reason the tradition going back to *McCulloch v. Maryland* puts such weight on reasoning about constitutional questions in light of constitutional purposes. We care about constitutional issues in order to further the commitments for which American government exists, and rightly so, for those commitments address matters that have a human significance deeper than any legal argument. Targeted killing is, I think, such a matter. I doubt anyone potentially within the sights of an armed drone would be comforted by the assurance that Congress has authorized the policy, or even (wrongly) that the president will provide him due process before ordering his death. What matters to the enemy targeted for a drone strike, as for the Americans whom that enemy threatens, has little to do with the Constitution's allocation of powers but everything to do with the Constitution's humane purposes. The law of the Constitution is vitally important, because unconstitutional federal action is lawless, and the deliberate but unconstitutional killing of a human being would be murder. Constitutional legitimacy is a legal necessity, but at the same time it is not a guarantee that the killing serves the Constitution's goals. Sound constitutional reasoning will point those who must decide such matters to look beyond

5. Felix Frankfurter, *Mr. Justice Holmes and the Constitution*, 41 HARV. L. REV. 121, 131 (1927) (quoted words reordered).

the Constitution to its goals. The final step when we think constitutionally is to think about the moral and political ends to which the Constitution is the means.

If it is unwise to engage in targeted killing because its net effect is to harm the security of the United States, as many think the current policy to be, then its lawfulness does not make it wise. If the president, or the American public, cannot honestly distinguish current policy from the sort of assassination that most of us would think immoral, then the U.S. government should renounce it, regardless of what the lawyers say. If it is wrong in certain circumstances to kill an al-Qaeda member who is a U.S. citizen, it is equally wrong (I think) to kill one who is not, whatever the limits of the Constitution's prohibitions. If U.S. drone strikes are imposing a reign of terror on people in some other part of the world, then we must stop because we are contradicting the deepest moral commitments of our political community. War is an ugly and brutal business, even war fought for a just cause and with the most scrupulous attention to legal norms. The limits on war that ultimately matter are those rooted in the humanity and decency of the society that goes to war to defend itself.

APPENDIX

Three Roads Not Taken

As the introduction noted, this book presupposes a particular understanding of how one ought to think constitutionally about issues in American government. Chapter Two presents that understanding by discussing two Supreme Court opinions, Chief Justice John Marshall's opinion for the Court in *McCulloch v. Maryland* and Justice Robert H. Jackson's concurrence in *Youngstown Sheet & Tube Co. v. Sawyer*, that are widely admired as classic examples of constitutional reasoning. Neither there nor elsewhere does the book offer a sustained justification for treating this understanding as normative, except insofar as a practical justification is implicit in my effort to think constitutionally about the issue of targeted killing along the lines laid out by Marshall and Jackson. An appendix is not a suitable vehicle for presenting a theoretical account of constitutional decision making, and I shall not attempt to do so here: For the purposes of this book, it is enough to observe that the constitutional approach I have adopted is (as my reliance on Marshall and *McCulloch* suggests) the traditional one followed since the beginning of U.S. constitutional history by most U.S. constitutional lawyers.

A leading constitutional lawyer of the present day, Charles Fried (Solicitor General of the United States under President Reagan and subsequently a Justice on the highest court of Massachusetts) has put the traditional Marshall/Jackson perspective well. The founders, Fried wrote, expected that

> the Constitution, like other legal texts, would be interpreted by men who were learned in the law, arguing cases and writing judgments

> in the way lawyers and judges had done for centuries in England and its colonies. Argument from precedent and by analogy would allow the Constitution to be applied to changing circumstances. . . . [P]recedent and analogy are the stuff of legal argument, and . . . legal argument is what moves the Court—or moves it when all involved are doing their work right. Certainly the Justices sometimes ignore or run over or through the arguments, but when this happens it is felt as unfair and wrong. And judges too must feel that precedent and analogy are the straw out of which their bricks are made. . . . This has been the texture of common lawyers' reasoning for centuries.[1]

As Fried put it more recently, "to act according to law [c]ertainly . . . implies knowledge of and judgment according to legal texts, but also precedents, traditions, legal principles, doctrines, and the customs and practices of courts, lawyers, and judges."[2] This book's presupposition as to constitutional reasoning is, in other words, that we can and should address the issue of targeted killing by drone using what are fundamentally the same intellectual tools that Chief Justice Marshall employed in addressing the issue of the national bank before the Court in *McCulloch*; the "methods of John Marshall's judiciary were—and are—entirely appropriate to" the tasks of legitimate constitutional thinking.[3]

Despite its historical status as (I believe) the constitutional mainstream,[4] this account of constitutional thinking is under serious challenge

1. CHARLES FRIED, ORDER AND LAW: ARGUING THE REAGAN REVOLUTION—A FIRSTHAND ACCOUNT 66 (1991).

2. Charles Fried, *On Judgment*, 15 LEWIS & CLARK L. REV. 1025, 1043 (2011).

3. FRIED, *supra* note 1, at 68. (The specific reference is to the "authority, powers, and methods" of the judiciary, but I do not think Fried would disagree with my extension of his words.) Neither Professor Fried nor I would deny the existence of change and development in the methods of reasoning and argument over the history of constitutional law, and of course there have been many substantive disagreements and changes of rule and doctrine. Both change and conflict are the marks of a healthy intellectual tradition, not of disarray nor of unwarranted departure from the tradition's original moorings. Justice Jackson's Steel Seizure opinion was, in my judgment, a faithful extension of the methods of *McCulloch*, but it was not a literalistic replication of them.

4. *Cf.* Frederick Schauer, *Is the Common Law Law?*, 77 CAL. L. REV. 455, 471 n.41 (1989) ("American constitutional adjudication in the Supreme Court seems a central case of common law methodology"). The modes of constitutional thought employed by lower court judges, competent lawyers filing briefs in any court, and executive branch legal advisors, are the same. *See, e.g.*, *The Constitutional Separation of Powers Between the President and Congress*, 20 Op. O.L.C. 124, 127 (1996) ("We believe that

at the present time; the criticism is particularly vigorous when the issues involved concern foreign affairs or national security. The reader may therefore find useful a brief critique of three alternative approaches to evaluating the constitutionality of an issue such as the lawfulness of drone strikes.

I. Radical Reform of the Law: Originalism

The first alternative I shall discuss is, in practical terms, the most significant competitor with the Marshall/Jackson/common law understanding of constitutional reasoning: originalism, the claim that the answers to constitutional law questions ought to be governed by what we can make out about the Constitution's original meaning.[5] "Originalism" in this contemporary and controversial sense is not, or not simply, acceptance of the importance of original meaning arguments in determining the sense to be given the text of a constitutional provision: *All* responsible constitutional lawyers, Marshall and Jackson and Fried included, have agreed on that.[6] Originalism, in the sense that the term carries in current debate, is the proposal that properly constructed original-meaning arguments have a priority over all other forms of constitutional argument, that at least presumptively and ordinarily an original-meaning argument persuasive in its own terms ("oh yeah, that *must* be what the Commerce Clause originally

the constitutional structure obligates the executive branch to adhere to settled judicial doctrine that limits executive and legislative power. . . . Our analyses are guided and, where there is a decision of the Court on point, governed by the Supreme Court's decisions on separation of powers.").

5. In referring to original "meaning" I am ignoring the many-sided debate among contemporary originalists about what counts as the authoritative original meaning (or understanding or intention—not even the word "meaning" is uncontroversial).

6. Walter Dellinger, another leading constitutional lawyer who, like Professor Fried, served as Solicitor General, calls this respect for the historicity of the constitutional text "moderate originalism." *See* H. Jefferson Powell, *Grand Visions in an Age of Conflict*, 115 Yale L.J. 2067, 2072 n.15 (2006) (Dellinger argues that "the existence of a written Constitution necessarily makes a moderate originalism an indispensable starting point for anything that can plausibly claim to be American constitutional law: A refusal to use original meaning to establish the starting point for the words the document uses would render the text infinitely manipulable."). "Moderate originalism" in Dellinger's sense is a starting point but not, as is true of

meant!") defeats other claims about constitutional interpretation. The meaning of the Constitution's spare language is to be fixed by an interpretive lodestar located in the past, beyond the reach of present-day preferences, and constitutional issues are to be resolved on the basis of historically determined textual assertions. This need not mean that constitutional lawyers can never entertain any other considerations at all—originalists differ on what weight (if any) to give Supreme Court precedent—but in principle contemporary originalism reduces legitimate constitutional argument to argument over the original meaning of textual provisions.

The scholarly literature on originalism, in the significant and controversial sense, is voluminous, but my basic objections to it can be simply put. First, despite frequent assertions to the contrary, originalism is not the historically original or historically dominant understanding of how constitutional decisions are to be reached. As Professor Fried has correctly observed, "[o]riginalism was not the original interpretive doctrine of the framers or of the framing generation."[7] The founding-era constitutionalist whose views were closest to those of a modern originalist was Thomas Jefferson, and as Chapter Two showed it was not he but his intellectual nemesis Marshall who decisively shaped the future of constitutional thought. Originalism is a proposal for revolutionary change, for an almost wholesale reformation of constitutional law—an aspect of the originalist project that is often overlooked, not least by originalism's proponents. If present-day originalists succeeded in persuading the rest of us, the forms in which American lawyers have historically made constitutional arguments would have to be severely curtailed.

The impact on constitutional law of replacing the Marshall tradition with modern-day originalism would therefore be dramatic and unpredictable in detail. Many of the Supreme Court's greatest decisions have given no sustained attention to original meaning at all: think of *Marbury, McCulloch, Gibbons, Strauder, Darby, Barnette, Brown, New York Times v. Sullivan, Reynolds v. Sims.*[8] In none of these cases (or a myriad others) did the Justices

the originalism debated in the law reviews, an outer limit on legitimate constitutional thinking.

7. Fried, *supra* note 1, at 66. "Founding-era lawyers did not view a constitution as a box the interpretive contents of which were fixed upon its drafting or adoption. Constitutional interpretation wasn't a strictly historical inquiry into what someone thought at some past point in time." H. Jefferson Powell, A Community Built on Words 65 (rev. ed. 2005). For examples demonstrating these claims, *see id.* at 48–50, 60–66, 85–100.

8. Marbury v. Madison, 5 U.S. (1 Cranch) 137 (1803) (Marshall, C.J.) (famous decision affirming the power of judicial review); McCulloch v. Maryland, 17 U.S. (4 Wheat.) 316 (1819) (anyone reading the footnotes will know this decision by now);

find it necessary to make serious originalist arguments, or impossible to reach a decision without doing so. All of these decisions would have to be deemed terribly wrong-headed (at least in their reasoning), and certainly should not be repeated, if originalism prevailed. The guidance that Supreme Court decisions should provide lawyers—and in particular lawyers advising the president and other executive branch policy makers—would therefore become quite unclear: Why should the president's lawyers follow the reasoning of a restrictive high Court precedent when a plausible original meaning argument points to the opposite conclusion? In the absence of a convincing argument that such drastic intellectual surgery is necessary, I think the American constitutional system should decline—and in the end will reject—the originalist proposal.

For the sake of argument, however, let us assume that sufficient cause exists for reducing constitutional law to the narrow range of arguments that "originalism" licenses. My second objection to the originalist proposal is that there is no obvious answer to the question of which of the many originalisms currently on offer we are to adopt.[9] Internecine conflict

Gibbons v. Ogden, 22 U.S. (9 Wheat.) 1 (1824) (Marshall, C.J.) (famous decision discussing the commerce powers of Congress); Strauder v. West Virginia, 100 U.S. 303 (1880) (seminal decision holding unconstitutional a racially exclusionary statute); United States v. Darby Lumber Co., 312 U.S. 100 (1941) (unanimous decision upholding broad federal regulatory powers); West Virginia Bd. of Educ. v. Barnette, 319 U.S. 624, 642 (1943) (Jackson, J.) ("If there is any fixed star in our constitutional constellation, it is that no official, high or petty, can prescribe what shall be orthodox in politics, nationalism, religion, or other matters of opinion or force citizens to confess by word or act their faith therein."); Brown v. Bd. of Educ., 347 U.S. 483 (1954) (seminal decision invalidating de jure racial segregation in public schools); New York Times v. Sullivan, 376 U.S. 254, 270 (1964) (First Amendment embodies "a profound national commitment to the principle that debate on public issues should be uninhibited, robust, and wide-open"); Reynolds v. Sims, 377 U.S. 533 (1964) (seminal one-person/one-vote decision). The list could be extended indefinitely.

9. Critics often point out the fissiparous tendencies of originalism.

> Some of the varieties [of originalism] include the following: It all began with conventional "intention-of-the-Framers" originalism. Then it became "intention-of-the-ratifiers" originalism. Of course, we also have "original-expected-applications" originalism... Then came "original-meaning" originalism, which was refined as "original-public-meaning" originalism (officially, this is now the position of Justice Scalia . . .). Justice Scalia himself distinguished "strong-medicine" or "bitter-pill" originalism from "faint-hearted" originalism. Then came "broad" originalism ... Now comes the "new originalism"... as distinguished from the "old originalism." Finally, we add "abstract" originalism ... Indeed, Mitchell Berman

among originalists over how properly to make originalist arguments is pervasive, and originalists generally are dogmatic about the obvious correctness of their preferred version. As a result, card-carrying originalists often reject one another's interpretive methods with as much vigor as they criticize non-originalists. Michael Stokes Paulsen, for example, insists that "no 'normative' argument is necessary to justify original-public-meaning textualism as the sole, legitimate, correct method of constitutional interpretation. . . . If what one is doing is 'constitutional interpretation,' original-meaning textualism is the only enterprise consistent with that description."[10] On the other hand, Walter Benn Michaels defends what he views as the original originalism on the ground that "you can't (coherently and non-arbitrarily) think of yourself as still doing textual interpretation as soon as you appeal to something beyond authorial intention—for example, the original public meaning."[11] For Professor Paulsen, Professor Michaels is in the grip of "the originalist-intentionalist fallacy," while Michaels thinks an "original-public-meaning approach" like Paulsen's is no more a legitimate originalist interpretation of the text than an overtly non-originalist argument.[12] And so it goes as one reads originalist literature, leaving even a willing convert wondering where to turn.

The problem is not simply that there are too many flavors of originalism to choose among. More fundamentally, the disputes among originalists demonstrate the fact that there is no basis in existing constitutional practice, or in assumptions generally shared among constitutionalists, that grounds the argument for any particular species of originalism as

> has distinguished seventy-two varieties of originalism in his tour de force Originalism Is Bunk.

James E. Fleming, *The Balkinization of Originalism*, 2012 U. ILL. L. REV. 669, 670–71 (2012) (citing Mitchell Berman, *Originalism Is Bunk*, 84 N.Y.U. L. REV. 1 (2009)).

10. Michael Stokes Paulsen, *Does the Constitution Prescribe Rules for Its Own Interpretation?*, 103 NW. U. L. REV. 857, 918 (2009). We need not pause over what Professor Paulsen means by "original-public-meaning textualism."

11. Walter Benn Michaels, *A Defense of Old Originalism*, 31 W. NEW ENG. L. REV. 21, 21 (2009).

12. *Compare* Vasan Kesavan & Michael Stokes Paulsen, *The Interpretive Force of the Constitution's Secret Drafting History*, 91 GEO. L.J. 1113 (2003), *and especially id.* at 1141 n.96 ("crude intentionalism") *and id.* at 1145 n.114 (intentionalism as a "fallacy") *with* Michaels, *supra* note 11, at 35 (original public meaning approach is "not a kind of textualism [and] it is not a kind of originalism either. . . . So we cannot really prefer [original public meaning] to non-originalism on the grounds that it is looking for what the Constitution originally meant; after all, the hypothetical plain meaner is just hypothetical.").

over against the rest. Those of us who defend the Marshall tradition as the proper approach to constitutional decision have just such a basis: The approach to constitutional thinking classically expressed in *McCulloch* simply is, as a historical matter, the tradition of reasoning that most constitutional lawyers have accepted and followed and that continues to dominate the actual practice of law in litigation, advice, and decision by lawyers and judges.[13] Lacking such a historical grounding, arguments for one form or another form of originalism regularly fall back on claims about the definition of their terms, which invite the response that definitional assertions have no persuasive power on their own: They gain purchase on the mind only if one has some good reason to accept the definition being proffered other than the reiteration in a louder voice of the circular or lexicographic claim that X is Y. Smart originalists like Professors Paulsen and Michaels know as much and offer reasons supporting their definitions,[14] but their (conflicting) reasons are contestable and contested

A possible response to my second objection is that the problem lies not in originalism but in the originalists, who have simply failed, as yet, to agree on a disciplined originalist methodology to replace the common law mindset that goes back to Chief Justice Marshall and that originalists agree on rejecting. We simply need to wait for agreement on the correct account of originalism. As a practical matter, this is quite unworkable: What are we to do in the meantime about constitutional questions that must be answered? Furthermore, why should we assume that there will ever be anything approaching consensus among lawyers and judges about which theory of originalism to follow?

The fatal problem with originalism, however is not simply that it isn't ready yet for prime time; its intellectual flaws are intrinsic. Originalism

13. The heightened prominence of original meaning arguments in constitutional decisions by the Supreme Court over the last few decades does not require any significant modification of this statement. Virtually by definition, such decisions involve issues that are extremely controversial under existing case law, or as to which there is not a great deal of precedent, and the Justices naturally and appropriately turn to historical arguments about the text's original meaning to resolve disputes when other methods do not. Even so, high Court constitutional decisions resting primarily on precedent and analogy employed in the traditional common law fashion greatly outnumber those centrally dependent on any form of originalist argument. I owe these points to Sarah Powell.

14. Respectively, that the Constitution's text ordains original public meaning interpretation, and that a resort to authorial intention is necessary even to establish the language and genre of a document. Space (and perhaps the reader's patience) will not permit exploration of the various forms of originalism or their warring rationales.

began, in part, with justified professional unhappiness over undisciplined decision making by the Supreme Court in the 1960s and 1970s, and as a critical slogan the rhetoric of fidelity to the Constitution's original meaning (understanding, intention, etc., etc.) has served a useful purpose, but it is not and cannot be a satisfactory account of how to think constitutionally for several reasons.[15] From a practical perspective, originalism is patently an academic exercise in the pejorative sense of that adjective. Few originalists of any stripe actually mean for constitutional lawyers not sitting on the Supreme Court (or in a professor's chair) to employ originalist methods to disregard Supreme Court decisions: Originalism is an academic theory that the *Justices* should employ in making their decisions, which decisions, it usually goes without saying, will bind the lower courts and other decision makers. Almost no one thinks that originalism should mean that every federal district judge in the country should enjoy the prerogative of ignoring a relevant Supreme Court decision because in his or her view it is inconsistent with original meaning. In an originalist world, therefore, it would only be in the absence of any relevant high Court precedent that other lawyers be free to adopt the originalist methodology that the Justices would be obligated to employ.

This leaves the question of how to bring the Court's decisions to bear when, as is usually the case, there is relevant precedent on the specific constitutional issue to be addressed. In the originalist world, lower court judges and others would need some intellectual means for relating binding precedent to novel questions—and the supposedly originalist legal culture would likely evolve a constitutional law of "[a]rgument from precedent and by analogy" for that purpose that would closely resemble what we have today.[16] Indeed, it is difficult to believe that an originalist Supreme Court itself would review every constitutional issue from the ground up, and to the extent that the Court applied some form of *stare decisis* (respect for precedent) to its decisions, and thus was obliged to determine when a

15. On the valid critical impulse lying at the historical roots of originalism, and its ineradicable flaws as a constructive program, *see* H. Jefferson Powell, *On Not Being "Not an Originalist,"* 7 U. St. Thomas L.J. 259 (2010). I have borrowed from that lecture, and from an essay supplementing the lecture, in summarizing the case against originalism in this appendix. *See also* H. Jefferson Powell, *Further Reflections on Not Being "Not an Originalist,"* 7 U. St. Thomas L.J. 288 (2010).

16. Civil law systems, to be sure, have intellectual means for stabilizing legal decision making across the system that differ significantly from the common law's tools of precedent and analogy, but it is implausible to expect that American lawyers would adopt a very different approach to legal thought in constitutional cases from the one they employ everywhere else.

decision governs a new set of facts and when it does not, the Court itself would be engaged in creating a common law form of constitutional law.

There are other difficulties with any form of originalism when it is offered not as a critical rhetoric but as a limitation on the legitimate forms of constitutional argument, but for present purposes we need not spend further time on the issue. Originalism, in the contemporary and contested sense, is not the original, or the traditional, or (most of the time) the contemporary mode of argument of American constitutional lawyers. The common law approach of Chief Justice Marshall and Justice Jackson remains, as it historically has been, the usual American constitutional practice, and I think that a book addressed to a difficult and important contemporary issue in American constitutional law need not pause to resolve the claim that American government and law should abandon that practice.

II. Skepticism about the Law: Political Constitutionalism

In the past two decades a lively academic debate has sprung up over the distinction between what are often termed "legal constitutionalism" and "popular" or "political constitutionalism."[17] Some of this scholarship is primarily historical and descriptive, and work in this vein is often extremely illuminating about the powerful shaping force that extrajudicial political actors and movements exert on the evolution of constitutional doctrine by the Supreme Court and sources of legal argument and doctrine. For some

17. *See, e.g.*, MARK TUSHNET, TAKING THE CONSTITUTION AWAY FROM THE COURTS (1999); LARRY D. KRAMER, THE PEOPLE THEMSELVES: POPULAR CONSTITUTIONALISM AND JUDICIAL REVIEW (2004); BARRY FRIEDMAN, THE WILL OF THE PEOPLE: HOW PUBLIC OPINION HAS INFLUENCED THE SUPREME COURT AND SHAPED THE MEANING OF THE CONSTITUTION (2009). The literature has grown immense, and I intend no disparagement of anyone else's contribution, but this is not the place for a literature review. The participants in this discussion do not universally employ the same terminology, but for convenience I shall employ Mark Tushnet's language of legal versus political constitutionalism. *See* Mark Tushnet, *Emergency Powers and Terrorism-Related Regulation Circa 2012: Perspectives on Prerogative Power in the United States* [hereinafter *Emergency Powers*], *in* EXTRA-LEGAL POWER AND LEGITIMACY: PERSPECTIVES ON PREROGATIVE 199 (Clement Fatovic & Benjamin A. Kleinerman eds., 2013) (describing his terms as "the established terminology" while expressing the view that "it might have been better had the concept [legal constitutionalism] been labeled *judicial* constitutionalism").

scholars, however, the distinction is normative. Professor Mark Tushnet gives a succinct definition of the contrast from a normative perspective.

> For the legal constitutionalist, constitutional restraints on government power and constitutional provisions organizing government power are law in the ordinary sense. Those subject to the constitution's requirements are *legally* obligated to comply with those requirements, not merely morally required to do so or urged to comply as a matter of political prudence.... Legal constitutionalism employs the usual way of enforcing legal obligations—a court order directing the actor to comply with the court's understanding of the constitution—to enforce compliance with the constitution's requirements.... Political constitutionalism... takes constitutional requirements no less seriously, but insists that they be enforced primarily by political rather than legal mechanisms. For political constitutionalists, actors subject to the constitution's requirements should comply with them because of political morality.... Not courts, but ordinary political mechanisms—elections, pressure from interest groups, media exposure—lead political actors to conform to a constitution's requirements.[18]

Needless to say, there is no consensus on exactly how to describe this contrast or about its implications: In particular, I think that some political constitutionalists would be uneasy with Tushnet's description of "political *morality*" as the basis of constitutional obligation.

There are important disagreements on how political constitutionalism as a normative practice or set of practices relates to legal constitutionalism and the institutional location of legal constitutionalism in judicial decision making. Is the primary point of the distinction that many constitutional debates go on in nonjudicial settings rather than in the courts, or that the American constitutional system includes debates of normative import that are conducted through modalities that litigators and judges do not (usually) employ?[19] Unsurprisingly, the long-debated issue of "judicial supremacy" takes on a new dimension when one thinks in terms of political constitutionalism. As Tushnet comments, "[m]ost versions of

18. Tushnet, *Emergency Powers*, *supra* note 17, at 199–200.

19. On these questions, the work of Professor Keith E. Whittington has been widely influential. *See, e.g.*, Keith E. Whittington, Constitutional Construction: Divided Powers and Constitutional Meaning (1999); Keith E. Whittington, Constitutional Interpretation: Textual Meaning, Original Intent, and Judicial Review (1999).

legal constitutionalism [treat] the [Supreme C]ourt's understanding [of the Constitution] as definitive"; in contrast, normative theorists such as Tushnet are skeptical about the extent to which participants in the practices of political constitutionalism should accept the high Court's case law as controlling.[20]

Whatever else the idea of political constitutionalism may do, it does suggest a basis for skepticism about the use of ordinary constitutional law reasoning to resolve the constitutionality of the Obama administration's targeted killing policy. The courts will probably not resolve the question of the policy's legality in any direct fashion because of justiciability limitations: Debate over constitutional issues must take place in nonjudicial settings and perhaps will take place only there. Furthermore, there are, as the reader knows, no Supreme Court decisions that squarely decide many of the constitutional issues we must address in determining the lawfulness of the killing of Anwar al-Awlaki. This is the sort of issue tailor-made for those who argue that political constitutionalism is both a necessity and a potential rival to the sort of constitutional law announced by the courts: A constitutional question that (let us assume) must be resolved outside the courts, and on which the usual sources of constitutional law argument are opaque, seems a clear candidate for resolution by some approach other than that of the legal constitutionalist.

In order to see why political constitutionalism skepticism is *not* justified, let us look at a particularly ambitious application of the political constitutionalism perspective, Professor Stephen M. Griffin's recent book, *Long Wars and the Constitution*.[21] Griffin's goal is to explain why we do not have a clear grasp of the constitutional dimension of post–World War II U.S. history, which has been characterized in his view by "a continuing series of policy disasters and constitutional crises" that are the result of unilateral presidential decisions to use American military power as a tool of foreign policy in addition to employing that power in circumstances plainly

20. Tushnet, *Emergency Powers*, *supra* note 17, at 199. On judicial supremacy from the viewpoint of political constitutionalism, *see id.* at 209–10.

21. I focus on Professor Griffin's book not because it is particularly flawed but for the opposite reason: Griffin's is the strongest and most elaborate argument of which I am aware for the political constitutionalism approach to constitutional issues involving foreign affairs and national security, although Griffin does not actually employ the term "political constitutionalism." *See* STEPHEN M. GRIFFIN, LONG WARS AND THE CONSTITUTION 4 (2013) ("One main theme of this book is that the Constitution can influence policy even when it is not enforced by the courts."); *id* at 11 (war powers are not "part of the legalized Constitution, those aspects of the Constitution that are typically enforced by the judiciary."). Griffin discusses targeted killing only very briefly. *Id.* at 253–54.

implicating the national security of the country. "Beginning with Truman, nearly all post-1945 presidents have claimed the unilateral power under Article II to initiate war, 'real' war, full-scale war."[22] The usual response of those opposed to such "presidentialist" claims of authority has been to invert this argument and claim for Congress sole power over decisions to use military force, but Griffin (though he leans in that direction) believes that this response is too narrow and too open to the "eminently defensible [presidentialist] policy rationale" that unilateral presidential action is justified because Congress patently lacks the institutional capacity to make foreign policy. The president is (or claims to be) the primacy formulator of U.S. foreign policy, and the use of military force is a crucial foreign policy tool in a world dominated first by the Cold War and then by the war against terror. "[T]he congressionalist critique has missed the real problem," which is precisely the constitutional claim that "military force is one tool among others" in the president's foreign policy toolkit and can be used without any substantive "interbranch deliberation." Congressional and popular acquiescence in presidential claims of unilateral authority to use military force have led to "the lack of a cycle of accountability" and "a recurrent and, indeed, tragic pattern of interbranch relations and presidential decisions."[23]

Professor Griffin disavows any ambition to prescribe in detail a constitutional solution to these problems, although it is clear that he thinks the creation of effective processes of interbranch collaboration is critical. "The most important task with respect to war powers is simply to understand the position we are in," and he describes the book as "more of an analytical history of presidential decisionmaking than a legal treatise."[24] What that history reveals, in his judgment, is that the original "constitutional order" with respect to foreign policy and national security was transformed after World War II by the actions of presidents and other executive branch officials, often with the approval of or at least acceptance by Congress and the public. The original constitutional order, a concept Griffin defines as a "relatively stable pattern[] of institutional interaction with respect to basic aspects of the Constitution such as powers and rights,"[25] required serious interbranch deliberation between Congress and the president on issues of foreign affairs and national security. While Griffin accepts, "in its broad outlines," the long-standing presidential claim to "executive leadership in foreign affairs [as] justified by the text and eighteenth-century history of the

22. The executive branch's lawyers, as we have seen, typically do not make such unguarded claims of unilateral power.

23. *Id.* at 7–10.

24. *Id.* at 9, 6.

25. *Id.* at 14.

Constitution," he insists that in the original constitutional order, "among the tools of diplomacy and national security strategy, war [was] special [and this made] approval by Congress obligatory before the nation goes to war."[26]

After 1945, or perhaps more specifically after President Truman's 1950 decision to go to war in Korea without obtaining congressional approval,[27] a new constitutional order came into being. The "solid base of the post-1945 constitutional order [lay] in concerns about threats to national security" posed by nuclear weapons and the Cold War, and brought again to the forefront by the events of 9/11 and its aftermath, and the contemporary constitutional order continues to seem unavoidable to many people because of "the obvious inability of Congress to contribute meaningfully to national security policy, at least without major structural reforms."[28] Despite its plausibility, however, the post-1945 constitutional order has repeatedly led to policy errors, some of them catastrophic. This is in part because the post-1945 order cannot be fully implemented due to the continuing presence of the original constitutional order's basic structure, and because post–World War II history has proven that the president and his advisors are systematically unable to make wise decisions about the use of military force when the executive uses force unilaterally and in pursuit of the president's unilaterally determined foreign policy goals.[29]

One might accept all of this—indeed, to a considerable extent I do—and still see a place for legal constitutionalist arguments in addressing issues about the exercise of military force. After all, even if the president ought to consult with Congress, regularly and substantively, the question whether a particular military action is lawful still must be asked and answered by the executive branch itself.[30] Perhaps the most obvious reason for doing so using the same methods of argument employed by the courts

26. *Id.* at 3.

27. *Id.* at 32 (referring to "what I call the '1950 thesis' . . . that Truman's 1950 Korea intervention marked a sharp break in our constitutional tradition"); *id.* at 272 ("starting with Korea presidential power was built on *legal* claims").

28. *Id.* at 268.

29. Griffin emphasizes his claim that the original constitutional order is "imprinted on the structure of government institutions," *id.* at 15, more than his arguments about the incapacity of the executive to make prudent decisions when acting on its own, but the latter theme is important too. *See, e.g., id.* at 256. There is a very powerful argument that his claims about executive incapacity are correct. *See generally* RICHARD M. PIOUS, WHY PRESIDENTS FAIL (2008).

30. And if the president and other policy makers do not ask, their subordinates will. *See* Jack Goldsmith, *The Irrelevance of Prerogative Power, and the Evils of Secret Legal Interpretation*, *in* EXTRA-LEGAL POWER AND LEGITIMACY: PERSPECTIVES ON PREROGATIVE 217–18 (Clement Fatovic & Benjamin A. Kleinerman eds., 2013)

is that executive branch decision makers expect their lawyers to do so. "Legal constitutionalism" is the established mode of constitutional argument, and political office holders, unlike academics, value familiarity over originality: The more the lawyers' defense of the policy makers' actions seems boringly orthodox and unremarkable, the better it is from the latter's perspective. As Professors Gabriella Blum and Philip B. Heymann put it in their insightful discussion of the role of the executive branch lawyer:

> When a president asks, in good faith, for advice on the law, he is almost always asking his lawyers what the courts are likely to do, or, in the absence of judicial precedent, what has been the practice and precedent in the executive branch. He will also want to know what most lawyers would think and say and how the legality of the action would be understood by other important groups. He is interested in the repercussions for himself and for the executive branch of appearing to behave illegally or without regard for the law. It is hard to imagine why a president, in seeking advice, would ever be interested in the personal, idiosyncratic judgments of the lawyers he chose.[31]

We may worry that the pressure to come up with answers the president likes will compromise the lawyers' objectivity, but that is only to say that giving good legal-constitutionalist advice may be personally difficult, not that it can't be done. Professor Griffin's critique of legal constitutionalism is more radical, however: He thinks that legal constitutionalism is essentially irrelevant to constitutional thinking about war, and his reasons for thinking this would, if persuasive, be a convincing reason for skepticism about the value of examining the constitutionality of targeted killing using traditional constitutional law arguments.

(discussing "the legal bureaucratization of national security and the attendant culture of sensitivity to legal authorization before action"); GRIFFIN, *supra* note 21, at 275 ("once CIA personnel began running sabotage operations against Cuba from Florida, they began to wonder whether such operations were truly, well, constitutional").

31. GABRIELLA BLUM & PHILIP B. HEYMANN, LAWS, OUTLAWS, AND TERRORISTS 50 (2010). *See generally id.* at 47–62. Executive practice and precedent are among the sources for traditional constitutional-law argument. *See* NLRB v. Noel Canning, 134 S. Ct. 2550, 2559 (2014) ("'[l]ong settled and established practice is a consideration of great weight in a proper interpretation of constitutional provisions' regulating the relationship between Congress and the President") (citation omitted); *id.* at 2571 (emphasizing, in relying on the constitutional views of an early attorney general, that "[n]early every subsequent Attorney General to consider the question throughout the Nation's history has thought the same").

Griffin's reasoning begins with the observation that "foreign affairs and war powers are in general outside the legalized Constitution. In the sphere of the legalized Constitution, the judiciary, guided by forms of legal argument influenced by the common law, is supreme." In contrast, "the norms of the nonlegalized sphere do not . . . have the same status as those in the legalized Constitution. Given that they are not (or not typically) enforced by the judiciary, they lack the certainty, stability, and authority of legal norms," since, after all, "[t]he legislative and executive branches swim in a sea of policy and politics. To be sure, they make law, but not *entrenched* constitutional law."[32] As a rough description of U.S. constitutional arrangements, this is fair enough (although I would not put it quite this way), but he draws from the description a normative conclusion: Ordinary constitutional law arguments, the sorts regularly employed by lawyers arguing cases within "the legalized Constitution," are inappropriate for addressing an issue located in the domain of political constitutionalism.

Professor Griffin makes this assertion repeatedly throughout the book. "The standard war powers debate [is] occupied with legalistic distinctions not suitable for a sphere of the Constitution unenforced by the Supreme Court"; "contemporary debate . . . is widely seen as a stalemate [because] both sides approach the debate as if it were occurring in a judicial forum inside the sphere of the legalized Constitution"; "[w]ar powers are not part of the legalized Constitution, and so a garden-variety doctrinal justification . . . cannot be fully satisfactory [nor would] a president . . . be satisfied with such arguments"; "this argument [over the meaning of the War Powers Resolution] was surely an example of a rather arcane dispute over statutory construction substituting for a discussion that should have engaged more meaningfully with constitutional values."[33] Griffin does explain at length or with complete clarity why he thinks ordinary constitutional law thinking is "not suitable" for the task of engaging "meaningfully with constitutional values" in questions involving the use of military force, but by teasing out why he makes this assumption, I think we can see why he is wrong, and we need not be overly worried by the political constitutionalists' skepticism about legal constitutionalism.

Professor Griffin's central rationale appears to be the fact that there are not very many Supreme Court cases addressing issues having to do with war powers. "American constitutional law," he writes, "is firmly based on the common law tradition. To oversimplify a bit, this means there is no law without precedents and there are very few meaningful precedents in the area of war powers. This g[ives] competent lawyers very little to work

32. *Id.* at 15–16.
33. *Id.* at 240, 27, 169, 250.

with."[34] As Chapter Two demonstrated, Griffin is entirely correct that ordinary constitutional law is, methodologically, a form of common law, but the understanding he suggests here of the capacities of common law thought is not merely oversimplified "a bit"—it is eviscerated to the point of plain error. If competent common lawyers were stymied by the absence of judicial precedents in addressing constitutional questions, then the entire enterprise of constitutional law, including that which lies within what Griffin calls "the sphere of the legalized Constitution," would never have gotten off the ground. John Marshall famously had very few constitutional precedents to work with, and was disinclined to cite the ones he had, and he seems to have little difficulty in coming to legal conclusions about constitutional issues. This was not simply a product of Marshall's unique genius or his role as a judge or his special location in constitutional history. Constitutional argument, often of the most sophisticated nature and indistinguishable in its legal character from the decisions of the Marshall Court, began in Congress with the first session of the First Congress, and in the executive branch with, at the latest, President Washington's 1790 consultation with James Madison, Chief Justice John Jay, and Secretary of State Jefferson over the Senate's advice and consent power with respect to treaties.[35]

Move forward two centuries and there is a lot more case law, but the contemporary Supreme Court still addresses constitutional issues about which there is little directly relevant case law. A common law constitutional method is indeed centrally concerned with cases, but it is concerned with them not as unexplained points of decision but as judgments that are the product of reasoning that takes into account the meaning of the text,

34. *Id.* at 39.

35. On the early congressional and executive-cabinet debates, *see* David P. Currie, The Constitution in Congress: The Federalist Period, 1789–1801 (1997), and David P. Currie, The Constitution in Congress: The Jeffersonians, 1801–1829 (2001). For Washington's consultation, *see* H. Jefferson Powell, The President's Authority over Foreign Affairs 41–47 (2002). Jefferson wrote a formal legal opinion in response to Washington's inquiry that was thoroughly "legalistic" in style and substance without any apparent embarrassment over the lack of judicial precedent. Marshall's discussions of Jefferson's 1791 opinion on the unconstitutionality of the national bank, and of Madison's 1800 legislative report on the unconstitutionality of the Alien and Sedition Acts, treated Jefferson's and Madison's reasoning as formidable legal arguments on major constitutional questions not at the time addressed by any cases. Marshall thought Jefferson, at any rate, wrong, but wrong because Jefferson's arguments were not as legally persuasive as the answers that Marshall, and Hamilton before him, could give, not because Jefferson made a category error in making constitutional law arguments when there was no case.

its historical origins and original meaning, the structure of the governmental institutions the text creates, the ethos of limited government that the text reflects, the history of political practice under the text, the impact on government, society, and the individual of different conclusions—in short a broad array of considerations that can still be debated even if there are few or no judicial precedents and indeed may never be any.[36] I think the Supreme Court has said considerably more that is relevant to issues involving the use of military force than Griffin acknowledges, but even if that were not the case, the absence of developed judicial doctrine is no reason to reject the applicability of ordinary constitutional law reasoning.

Professor Griffin's other reasons for banning legal constitutionalism from the resolution of war-related issues are corollaries, I think, of his crabbed view of ordinary constitutional law's methodological capacities. "Contrary to what originalists may think," he writes, "there is no way to generate meaningful doctrine from the original constitutional order to answer every contemporary military contingency."[37] This is a striking *ipse dixit*, but that is all it is. If the original constitutional order required the president to obtain congressional approval for all uses of military force except in response to an actual military attack on the United States (a not implausible but I think erroneous reading), then we would need to develop a reasoned pattern of analysis to determine what counts as "an actual military attack on the United States."[38] I do not think doing so would be significantly more difficult than the generation of meaningful doctrine from the

36. This is not to deny that constitutional lawyers prefer to have judicial and more specifically U.S. Supreme Court decisions with which to work. Over twenty years ago Professor Peter J. Spiro wrote that in light of the "imprecision" of the constitutional text, the "longstanding refusal by the courts to act as referee," and the absence of "effective or unquestioned statutory disposition," the question of the president's authority to initiate hostilities is troubling to lawyers. "This want of the usual instruments of interpretation tends to distress lawyers, who are most comfortable when using the powers of analogy to puzzle through a set hierarchy of self-contained statutes and judicial opinions." Peter J. Spiro, *War Powers and the Sirens of Formalism*, 68 N.Y.U. L. Rev. 1338, 1339 (1993). Spiro was sharply critical of what he termed "refined parchment solutions" to war powers issues, but on the ground that an excessively textual approach to the issues ignores the greater persuasiveness of legal arguments based on structure and political practice: Formalism, ironically, "promotes the perception that war powers is [*sic*] a matter of politics, not law [and that] in turn undermines the efficacy of norms established by history and important to our constitutional order." *Id.* at 1355, 1341.

37. Griffin, *supra* note 21, at 250.

38. The principle I state in the text is, at least roughly, Professor Griffin's own view. *See id.* at 97. The principle actually to be drawn from the original constitutional order is rather more complex, in my judgment, but I have used this to-me

first amendment's laconic prohibition on "law . . . abridging the freedom of speech." In both instances, one must bring analogy, and the identification of clear examples of what is "an actual military attack" or "an abridgment of freedom of speech," to bear on the question, and the full development of doctrine on the scope of "actual attack" would take time and multiple examples of specific issues addressed and satisfactorily resolved, just as has been true with "freedom of speech." The difficulties are true of constitutional doctrine on any complex issue, within or without Griffin's "legalized Constitution"; all that this tells us is that sound constitutional law reasoning is not a matter of deduction in the sense that we might solve a problem in formal logic.[39]

Professor Griffin also expresses concern over the manner in which legal constitutionalist reasoning employs arguments from constitutional history. "Despite the lack of legal doctrine, scholars studying war powers tend to . . . focus on instances of conflict between the executive and legislative branches and treat them as if they were cases. Scholars using this approach are often trying to answer the abstract question of how the Constitution allocates war powers in any situation involving the use of military force."[40] Since Griffin himself believes that "[h]istorical analogies have a special importance in the realm of the nonlegalized Constitution," it is not entirely clear what his concern is. Perhaps it is only that in his view post-1945 presidentialists have *misused* historical examples to justify presidential actions on the basis of "a kind of deliberately constructed tradition . . . never based

faulty principle for ease in making a point that is any case the same: Professor Griffin's certainty that the modes of traditional constitutional thought cannot generate doctrine that answers specific present-day questions is mere dogma, and erroneous dogma at that.

39. Professor Griffin does not address directly traditional constitutional law's incorporation of several other common law modalities of argument. The omission is rather ironic since Griffin makes extensive use of structural, prudential, and constitutional-ethos based arguments, particularly in supporting his persuasive claim that the Constitution presupposes and indeed requires interbranch deliberation and a cycle of executive accountability with respect to foreign affairs and national security decisions. On these forms of argument, *see* the classic discussions in PHILIP BOBBITT, CONSTITUTIONAL FATE (1982) and CONSTITUTIONAL INTERPRETATION (1991). As with his dismissal of mainstream original meaning arguments, this seems to reflect Griffin's surprisingly narrow and erroneous view of the scope and range of traditional constitutional thought. There is no obvious reason why able legal constitutionalists should be any less capable than Griffin himself of drawing normative inferences from observations about governmental structure, the consequences of contrasting decisions, or the overall shape of the American constitutional enterprise.

40. *Id.* at 27. The language I quote below about "hypothetical cases" also comes from this passage.

on systematic and deeply contextual accounts of the past."[41] I have no doubt that has sometimes been true, but it is an accusation of poor reasoning, for which the remedy is better reasoning not an abdication of the enterprise. Griffin's description of political practice arguments as a misguided attempt "to answer [an] abstract question" is quite tendentious—one can describe *any* attempt to state the law on any subject outside the context of a specific legal question as necessarily involving a "focus on hypothetical cases" either explicitly or by implication. Professor Griffin's repeated assertions about the constitutional need for "a cycle of accountability" and "interbranch deliberation" are abstractly stated conclusions he has drawn from the historical instances of presidential unilateralism he discusses, and he clearly and quite properly wishes the reader to imagine hypothetical cases in which a prudent executive and an engaged Congress followed his admonitions.

Professor Griffin's attempt to excommunicate ordinary constitutional law reasoning from use in addressing the constitutional issues raised by the use of military force gives us no valid reason for skepticism about the applicability of that reasoning to these issues: His attack is on a straw man, not on the form of constitutional thinking presented in Chapter Two, which is as a factual matter the approach to constitutional issues that competent and conscientious lawyers in government would employ in addressing a question such as the validity of the targeted killing policy.

III. Nihilism about the Law: Prerogative Rather than Authority

As anyone who teaches introductory constitutional law in a law school knows, nothing is easier to adopt than the cynical assumptions that constitutional law has no independent substantive content, that constitutional arguments are window dressing for ideological conclusions reached on other grounds, and that the Supreme Court is a group of crafty politicians in black robes.[42] Many students come to the class already convinced by this

41. *Id.* at 205

42. The suspicion is very old. *See, e.g.*, Thomas Jefferson, Letter to Thomas Ritchie (Dec. 25, 1820), http://www.let.rug.nl/usa/presidents/thomas-jefferson/letters-of-thomas-jefferson/jefl263.php ("An opinion is huddled up in conclave, perhaps by a majority of one, delivered as if unanimous, and with the silent acquiescence of lazy or timid associates, by a crafty chief judge, who sophisticates the law to his own mind, by the turn of his own reasoning.").

story, and if the teacher is incautious, the Justices themselves will appear to confirm its truth. It is doubtless unnecessary to elaborate on the experiences of the much smaller set of academic lawyers who attempt to teach executive branch constitutional law opinions: The president, after all, *is* a politician, and the widespread assumption that "politics" and "law" are antithetical makes it a heavy lift to persuade students to entertain the possibility that executive branch law is anything other than executive branch policy preferences dressed up in legal garb.

From the popular-cynical perspective, then, worrying about lawyers' arguments over the constitutionality of targeted killing is self-evidently unnecessary or even delusional: Of course such arguments are of no moment as long as the question doesn't get before a court, and the same is probably true even then.[43] This sort of cynicism is as uninteresting as it is irrefutable. But there are arguments by serious scholars that come to much the same conclusion, and if they are right, then the attempt to work out what the executive branch's constitutional reasoning about the targeted killing policy *should be* is a parlor game for academics. The most prominent recent argument to this effect is in a book by Professors Eric A. Posner and Adrian Vermeule, *The Executive Unbound*, and I shall address all-out skepticism about the practical significance of constitutional law for executive branch decision making by considering their book.[44]

Posner and Vermeule describe their "main critical thesis" as the claim that what they call "liberal legalism" has failed with respect to the presidency. By liberal legalism they mean a viewpoint that they attribute to mainstream legal thinkers and trace back to the original (Madisonian) constitutional design: The president is limited by the formal legal boundaries on his power and checked by a vigorous, proactive Congress. In fact,

43. The federal courts have turned aside as nonjusticiable attempts to secure substantive review of executive actions in the war on terror. *See* al-Aulaqi v. Obama, 727 F. Supp. 2d 1 (D.D.C. 2010) (granting on justiciability grounds a motion to dismiss an action brought by al-Awlaki's father to enjoin the targeted killing of his son). *See also* El-Shifa Pharmaceutical Industries Co. v. United States, 607 F.3d 836 (D.C. Cir. 2010) (holding that the political question doctrine barred review of the legality of a cruise missile attack on a Sudanese manufacturing plant), *cert. denied*, 131 S. Ct. 997 (2011). I do not mean to imply that these decisions were reached on any basis other than good faith legal reasoning. The popular-cynic might respond that justiciability doctrines supply convenient legal excuses for avoiding the adjudication of controversies in which federal judges would rather not be involved.

44. *See* Eric A. Posner & Adrian Vermeule, The Executive Unbound: After the Madisonian Republic (2010).

however, "liberal legalism has proven unable to generate meaningful constraints on the executive."

> Liberal legalism, in short, has proven unable to reconcile the administrative state with the Madisonian origins of American government. The constitutional framework and the separation-of-powers system generate only weak and defeasible constraints on executive action. . . . [T]here is an enormous world of constitutional problems that the judges rarely touch, involving the structure and function of government. . . . [A]s a legal mechanism for ensuring that the executive remains within the bounds of law, [congressional] oversight is largely a failure. . . . Under the civilizing force of hypocrisy, [executive branch] actors will incur a cost if they act too opportunistically, [b]ut that does not require, indeed it implicitly denies, that [executive] actors have internalized [constitutional practice] as legally obligatory.[45]

In short, we live "in a world of little constitutional constraint," and Congress's attempts to create statutory "instruments of executive constraint . . . having been tried, they have been found wanting."[46] This doesn't mean that the United States is in the grips of an "imperial presidency," Posner and Vermeule reassure us, because although "law does little to constrain the modern executive, contrary to liberal legalism's hopes, . . . politics and public opinion do constrain the modern executive, contrary to liberal legalism's fears."[47] "Tyrannophobia" is a cultural neurosis, justified by the false belief, inherited from our Madisonian past, "that the only possible constraints on the executive are de jure constraints." Once we recognize that power of the political constraints on the executive, we spend no more time worrying about the fact that "Madisonian separation of powers is obsolete."[48]

The argument in *The Executive Unbound* against the meaningfulness of law (and specifically constitutional law) in constraining the president verges at times on being so overheated that its extreme rhetoric threatens to obscure its more sober points.[49] It would be a mistake, however, to

45. *Id.* at 7, 10, 65, 75.

46. *Id.* at 89.

47. *Id.* at 15.

48. *Id.* at 204.

49. Professor Richard M. Pious began a review of *The Executive Unbound* with a paragraph summarizing, briefly but fairly, the book's thesis and then observing, "*I am not making this up.*" Richard M. Pious, Book Review, 41 PRES. STUD. Q. 864, 864 (2011). *The Executive Unbound* has been subject to a drum beat of negative

ignore the challenge that Posner and Vermeule rightly put to any easy presumption that constitutional law is an effective constraint on presidential decision making in areas, of which targeted killing is surely one, where judicial review is likely to be highly deferential or completely nonexistent. One need not subscribe without reservation to the proposition that rational choice theory ("the preferred Posner-Vermeule analytic toolkit"[50]) resolves all puzzles to agree with them that there are obvious reasons for a president to avoid, to the maximum extent possible, any roadblocks to his or her policy goals thrown up by the law. Nor is there any real doubt that the modern presidency has a great degree of initiative in policy making, or that recent Congresses tend to be reactive if not actually passive in the face of executive action, or that smart lawyers can devise colorable arguments out of the Constitution and many statutes to justify a very broad range of executive policy choices.

Posner and Vermeule are right to raise the skeptical question whether constitutional reasoning limits presidential decision making in a significant fashion, but their ostensible preference for an executive that makes no claim to be bound by law renders the question unhelpfully theoretical if taken literally. To be sure, their book fits into an extensive recent literature re-examining the idea of executive "prerogative," the historical term for the asserted power to act in exigent circumstances *extra legem* or even *contra legem*, but that notion probably has no "continuing practical relevance

commentary. A common thread running through many of the critiques is the observation that Posner and Vermeule tend "to substitute rational speculation for closer examination into historical sources," Mark A. Graber, Book Review, 11 Perspectives on Politics 1157, 1159 (2013), with the ironic result that "a book that begins by taking liberal legalism to task for being disconnected from reality . . . base[s] so much of its discussion on abstractions." Chris Edelson, Book Review, 127 Pol. Sci. Q. 311, 311 (2012). It's not even clear at times how seriously the authors mean us to take their broadest claims. *Compare The Executive Unbound*, at 75 (executive actors do not internalize political precedents as obligatory) *with id.* at 72 (political precedents "tend to ossify into institutional routines and individual habits, and may even become internalized by actors who develop a sense of legal obligation to follow the precedent"). There is an air of *épater la bourgeoisie* about much of the book.

50. Peter M. Shane, Madisonianism Misunderstood: A Reply to Posner and Vermeule (Apr. 8, 2011), http://www.acslaw.org/acsblog/madisonianism-misunderstood-a-reply-to-posner-and-vermeule. I am not concerned with the question of how plausible Posner and Vermeule's use of rational choice theory is within its own terms. *See* Harvey Mansfield, Is the Imperial Presidency Inevitable? (Mar. 11, 2011), http://www.nytimes.com/2011/03/13/books/review/book-review-the-executive-unbound-by-eric-a-posner-and-adrian-vermeule.html?pagewanted=all&_r=0 (criticizing their "incongruous reliance on the rational actors of game theory").

to American constitutionalism.... [T]he Jefferson-Lincoln conception of prerogative power—most notably, executive action in open defiance of law or legal authority—is no longer part of a president's justificatory tool kit."[51] For reasons that Jack Goldsmith (a senior executive branch legal advisor to President George W. Bush) in particular has laid out persuasively, it would be extremely surprising for a modern president even to consider openly using "prerogative" power in its historical sense.[52] But we need only adjust Posner and Vermeule's skepticism slightly in order to state it as a plausible course of action.

One of the leading students of the presidency, Richard M. Pious, has long warned that the executive branch finds constitutional rhetoric to be empowering rather than constraining: "[T]he irreducible core of governmental power rests ... with executive officials' substitution of contestable constitutional prerogatives for congressional authorization." The strategic use of constitutional reasoning by presidents and their lawyers thus underwrites "the development of 'prerogative governance' as a political style" that frees presidents from "checks and balances [and] statutory frameworks requiring interbranch policy collaboration."[53] As revised by Pious, then, the Posner and Vermeule skepticism can be put this way: An attempt to work out the best constitutional reasoning about the problem of targeted killing has no purchase on the real world because in the absence of judicial review a president will adopt the reasoning only if it provides plausible "claims of prerogative power" that can serve as "institutional resources to advance [the administration's] goals,"[54] and will disregard the argument if it creates

51. Goldsmith, *supra* note 30, at 214.

52. In addition to the essay just cited, *see generally* JACK GOLDSMITH, THE TERROR PRESIDENCY: LAW AND JUDGMENT INSIDE THE BUSH ADMINISTRATION (2007).

53. Richard M. Pious, *Public Law and the "Executive" Constitution*, *in* EXECUTING THE CONSTITUTION: PUTTING THE PRESIDENT BACK INTO THE CONSTITUTION 11, 13, 27–28 (2006). *See also* Henry P. Monaghan, *The Protective Power of the Presidency*, 93 COLUM. L. REV. 1, 2 (1993) (describing a 1979 book by Pious as "a rare dissent as to the importance of constitutional law"). Professor Monaghan's classic article is an exploration into the extent to which U.S. constitutional law legitimately accommodates the felt need for executive action *extra legem* in emergencies or other exigent circumstances. *See* Monaghan, *id.* at 33 n.157 ("for me the question is whether in legal analysis the national government's emergency power is extra-constitutional").

54. Pious, *supra* note 53, at 13. Pious has analyzed the Obama administration as relying heavily on what Pious calls the assertion of "soft prerogative"—the administration usually advances legal arguments for its policies that ostensibly defer to congressional authority, for example by relying on statutes rather than Article II of the Constitution as the source of the president's power, while construing the statutes to maximize presidential discretion, and noting without any fanfare the possibility of

difficulties. One can usefully write briefs for future presidential actions (or against them), if one is so inclined, but worrying over the correct analysis, even if in some jurisprudential universe the word "correct" is meaningful, is beside the point as a practical matter.

There is, to be sure, nothing novel about advancing claims of legal power for strategic political purposes. Justice Jackson referred in his *Youngstown* concurrence to result-determined constitutional claims about presidential power as a "persuasive dialectical weapon in political controversy" and contrasted such "self-serving press statements" with the kind of serious constitutional arguments that the president's lawyers would endorse in giving him actual, confidential advice or in representing the administration in court."[55] If Posner and Vermeule's skepticism were well-founded, constitutional inquiry into presidential authority in areas unlikely to be subject to judicial review would be a study in what is likely to move public or congressional opinion in a political controversy. However, I think that there is a persuasive response to the important question they raise.

The starting point for seeing the flaws in Posner and Vermeule's portrait of a president unbound by law is to recall that the presidents and their advisors can and in fact usually do prefer to act within legal bounds as a matter of conscious choice. In the middle of World War II, Justice Jackson asserted that the reason the exercise of extraordinary powers during wartime had not, up to that point, "overwhelmed our constitutional system," was that "[e]ach crisis so far has been handled by those in Executive authority trained in the tradition of the law [whose] fidelity to the Constitution on the whole does not suffer by comparison with that of the Court."[56]

invoking Article II in appropriate circumstances. *See* Richard M. Pious, *Prerogative Power in the Obama Administration: Continuity and Change in the War on Terrorism*, 41 PRES. STUD. Q. 263, 284 (2011) ("most of the [Obama administration] claims to prerogative powers are buried deep inside the legal briefs and motions filed by administration lawyers").

55. *See Youngstown*, 343 U.S. at 647. Immediately before the lines I quote in the text, Jackson referred to the "pressures within and without an administration for presidential action beyond that supported by those whose responsibility it is to defend his actions in court." For a reference to the careful private constitutional advice that Jackson believed the executive branch's lawyers traditionally provide, *see id.* at 645 ("advice to the President in specific matters [concerning the scope of the commander in chief power] usually has carried overtones that powers, even under this head, are measured by [limiting considerations]. Even then, heed has been taken of any efforts of Congress to negative his authority.").

56. Jackson, unpublished concurrence in *Hirabayashi v. United States*, 320 U.S. 81 (1943), printed in Dennis J. Hutchinson, *The "Achilles Heel" of the Constitution: Justice Jackson and the Japanese Exclusion Cases*, 2002 SUP. CT. REV. 455, 473.

There is no reason to doubt that this remains the case: In fact, as Professor Goldsmith as noted, "the hyper-legalization of war and national security" in recent decades has, if anything, made executive branch officials even more conscious and concerned about the legal limitations on their actions.[57] There are many sources of this strong tendency on the part of presidents and their advisors to seek consciously to remain within the law: personal commitment to constitutional values, the fact that belief in the law's normative force and fear of being accused of breaking the law influence behavior, and social pressure within executive branch offices and agencies.[58] The lawyers who provide advice to the president and the officials on whom the president must rely to carry out his or her policies are alike in that most of the time they are unwilling to advise or act in contradiction to what they perceive the law to be.[59] Posner and Vermeule's assumption to the contrary has neither an empirical nor a historical basis.

The model of human behavior that *The Executive Unbound* superimposes on the process of executive branch constitutional decision making fails to take into account the conscious reasons presidents and their advisors have for respecting legal limits. The book also neglects for the most part the

57. Goldsmith, *supra* note 30, at 220. It is fair to observe that with respect to one, extremely important constitutional issue, the president's constitutional authority to make use of military force in the absence of congressional authorization, presidents have become significantly bolder since World War II. *See id.* at 224 (noting that "there is still fundamental debate and disagreement over whether and when the president" may act unilaterally). Whether this reflects a fundamental change in the executive branch's understanding of the president's authority, *see generally* GRIFFIN, LONG WARS AND THE CONSTITUTION, or the executive's continued adherence to constitutional norms dating back in principle to the early Republic, *see generally* POWELL, THE PRESIDENT'S AUTHORITY OVER FOREIGN AFFAIRS, it is clear that the rhetoric of presidential unilateralism has become more extreme in the past two-thirds of a century.

58. As Benjamin Kleinerman put it in reviewing Posner and Vermeule, "[w]hile realizing that the law does not enforce itself automatically ... Madisonian realism also realizes that the law transforms politics." Benjamin Kleinerman, *The Madisonian Constitution: Rightly Understood*, 90 TEX. L. REV. 943, 968 (2012).

59. *See, e.g.*, David Cole, Are We Stuck with the Imperial Presidency? (June 7, 2012), http://www.nybooks.com/articles/archives/2012/jun/07/are-we-stuck-imperial-presidency/ ("In my experience, lawyers for the executive branch generally take legal limits seriously."); Goldsmith, *supra* note 30, at 217 ("executive branch national security officials have become deeply acculturated and sensitive to legal authorities and restrictions. They will not act in defiance of law, even if the president orders them to or says it is acceptable to do so."); GRIFFIN, *supra* note 21, at 275 ("Most citizens and certainly most government personnel know that the Constitution is a law that the government must respect").

mindset with which executive branch officials are overwhelmingly likely to approach a constitutional question. It is not simply that they have conscious reasons, disinterested or self-serving, for finding plausible answers. Beyond that, most of them will consider the question from a perspective deeply shaped already by the role of law in American life and society. As Mark A. Graber puts it, the Constitution "constitutes politics."

> [C]itizens do not develop values and interests independently from constitutional norms. Legal norms influence the values citizens hold and the interests they pursue in politics.... Such persons, when they vote or hold public office, often have no desire to act contrary to constitutional norms.... Most governing officials and citizens... internalize basic constitutional values.[60]

A president's legal advisors, and sometimes the president himself or other high policy makers are "trained in the tradition of the law" (Jackson), which strengthens the general American predisposition to prefer decisions that are seen as respecting the Constitution.[61] Professor Pious is certainly right to insist on the possibility, and often enough the reality, of sheerly strategic invocations of constitutional law; at the same time, "much constitutionally constrained behavior, including the practice of providing legal justifications for decisions, cannot easily be explained away as strategic efforts to secure desired policies."[62]

It is wildly inaccurate to view human beings as nothing more than the rational choice actors Posner and Vermeule imagine them to be, but there are limitations on even the idealistic decision maker. The construction and evaluation of constitutional arguments are affected by human cognitive limitations, including the distorting effect of idiosyncratic constitutional or ideological beliefs, and of motivated reasoning in general, and against such dangers a sincere commitment to obeying the law is no perfect safeguard against error. "The danger is that legal analysis can be self-serving and mistaken, and that the executive will as a result act extra-legally."[63] There is nothing new about this recognition that cognitive limitations pose a threat to constitutional law, which is a social practice that depends on the exercise of reason: Chief Justice John Marshall wrote in 1805 that "[t]he

60. Mark A. Graber, A New Introduction to American Constitutionalism 214–15, 242 (2013).

61. "Lawyers and other constitutional decision-makers are particularly prone to think almost exclusively in constitutional categories." *Id.* at 236.

62. *Id.* at 217.

63. Goldsmith, *supra* note 30, at 225.

judgment is so much influenced by the wishes, the affections, and the general theories of those by whom any political proposition is decided, that a contrariety of opinion on this great constitutional question [the validity of the national bank] ought to excite no surprise."[64]

Marshall did not conclude from this fact, however, that the quest to make sound constitutional judgments is hopeless, nor should we. The difficulty can be addressed in part simply by recognizing the possibility of a "contrariety of opinion" and one's own likely biases. Nor is it unrealistic to expect the executive to engage in this sort of critical self-awareness. Even in the confidential back-and-forth between presidents and their lawyers there will ordinarily be an imagined and potentially hostile audience whom both president and advisor have in view in evaluating different lines of constitutional argument. This check on unintended error can break down, of course, but it is not insubstantial. It would be naive to deny that motivated reasoning and its group-related cognates render illusory any idea that the president and his or her advisors will make difficult constitutional decisions on the basis of a sheerly objective evaluation of the arguments pro and con—but then that assumption would be equally naive if applied to Supreme Court Justices or members of Congress, or even to oneself and one's colleagues.[65] What is important for present purposes is the fact that while a sophisticated understanding of how human beings make complex normative decisions may complicate our understanding of constitutional law as a practice of reasoned decision making, such an understanding greatly weakens the force of Posner and Vermeule's skepticism.

The claim that constitutional law is insignificant in executive decisions such as those related to targeted killing, furthermore, is plausible only if one ignores or is unaware of significant features of the actual executive branch of the U.S. government and the ways it functions. Skeptics such as Posner and Vermeule often seem to assume as a background to their discussion of details a simple and stylized image of the executive as a single figure possessed of unity of purpose, an unerring sense of the national interest (particularly with respect to national security), and the unimpeded capacity to control the actions of his subordinates. For such a Titanic figure, constitutional law could only be irrelevant if favorable or a problem to be swept away if not. It is a dramatic if not very democratic picture, but it is not a portrait of the U.S. executive branch. There is indeed, a single president, but the executive branch is in fact a vast, heavily bureaucratized system of

64. 4 John Marshall, Life of George Washington 243 (reprint ed. 1983) (1805).

65. *See* Dan M. Kahan, Foreword, *Neutral Principles, Motivated Cognition, and Some Problems for Constitutional Law*, 125 Harv. L. Rev. 1 (2011).

departments, agencies, and offices, with many different individuals exercising significant authority and doing so much of the time at least partly at cross-purposes with one another and with the president, who has come in recent decades to preside over his or her own mini–executive branch within the Executive Office of the President (EOP). The EOP's purpose, and to some extent its actual role, is to enhance the president's ability to shape what the "big" executive branch does, but EOP too has become bureaucratized and Balkanized. In such a complex network, legal constructs such as "the unitary executive" and the removal power as a significant instrument of presidential control are remote from the truth, which is much messier and less rational. Posner and Vermeule imagine a twenty-first-century government in which centralized and unitary executive power has triumphed; the reality is more one of dispersed and often poorly coordinated administrative power, with the president frequently grasping at straws in his efforts to impose a modicum of overall purpose.

In this world, far from being a problem, the law, constitutional law in particular, is often one of the president's most potent tools. The "official" constitutional law of separation of powers, both the judicial decisions found in the United States Reports and the executive's own elaboration of doctrine, is tilted quite far in the president's favor.[66] The president's lawyers assist through the ordinary roles of legal counsel in reducing to more manageable size a problem by identifying options, limitations, and loci of decision-making authority, and they (and their constitutional opinions) are valuable resources in the president's unending struggle to gain some mastery over "his" or "her" branch of government.[67] At the same time,

66. *See, e.g., The Constitutional Separation of Powers Between the President and Congress*, 20 Op. O.L.C. 124 (1996). Whether this tilt is the best reading of constitutional text, structure, and history is debatable, but we need not address that issue here.

67. I should note that the indirect power of rendering legal advice in the executive branch is itself dispersed, with the most prominent locations being the Justice Department's Office of Legal Counsel (OLC, exercising the attorney general's constitutional and statutory duties to give advice and thus representing a tradition dating back to the administration of George Washington), L (the Office of the State Department Legal Adviser, and laying claim to special expertise on treaties and other international law matters), and the White House Counsel's Office (with the most strategic location and for that reason the least institutional capacity for detached analysis). OLC has the formal authority to adopt legal conclusions for the executive branch, unless it is overruled by its own Justice Department superiors or the president, but the existence of multiple sources of counsel makes it possible for presidents to shop around for advice—and critics of the president to assail him for doing so.

the work of the lawyers ensconced in the departments and administrative agencies supports the goal of administration coherence by heading off legal problems or referring difficult constitutional issues to lawyers more closely connected to the president for resolution. The realities of twenty-first-century government thus undergird the importance of constitutional law in the executive: Presidents rely on their lawyers' valuable work in other respects and become accustomed to soliciting legal advice as one aspect of decision making, and the dispersal of actual governmental authority throughout a vast administrative apparatus, peppered with lawyers socialized to think in constitutional terms and to insist on constitutional analyses, makes "administrative constitutionalism . . . a necessary corollary of the reality of administrative government."[68] Posner and Vermeule's story of the president breaking free of the entanglements of constitutional and legal constraints is almost the reverse of the truth in practical, organizational terms. Their dismissal of law as irrelevant to executive branch decision making, far from resting on a hard-headed realism about the world, is as naïve as it is nihilistic.

* * *

There is a common thread running through the originalist proposal to eviscerate the traditional range of legitimate constitutional arguments, the political-constitutionalist call for a "nonlegalized Constitution," and the overt constitutional nihilism of a book like *The Executive Unbound*. The originalist, the political constitutionalist, and the nihilist share a strong disbelief in the value of the legal tradition that John Marshall played a central role in creating: It is too freewheeling for the first, too unimaginative for the second, too irrelevant for the last. There are, I have argued, specific but equally fatal flaws in the views of each, but their central and common error is their unwillingness or inability to recognize the intellectual and moral force of the Marshallian tradition. The great constitutional lawyer Charles Black once wrote that law is an art, and that as an art it demands both freedom to give "full scope to such intuitive factors as one possesses" and fidelity to the tradition in which one works: "[R]esponsibility to reason, even to technical reason, is the soul of the art of law."[69] To our

68. Gillian Metzger, *Administrative Constitutionalism*, 91 TEX. L. REV. 1897, 1924 (2013). Professor Metzger emphasizes the normative aspect of her observation while my focus is more descriptive, but her discussion seems to me correct both normatively and descriptively. *See id.* at 1925 ("in an administrative world, administrative agencies must become a locus of independent constitutional enforcement to do justice to the principle of constitutionally constrained government").

69. Charles L. Black, Jr., *Law as an Art*, *in* BLACK, THE HUMANE IMAGINATION 26 (1986).

"postmodern" sensibilities, such claims for constitutional law may seem either woolly-headed or disingenuous, but Marshall and Black and Robert Jackson knew better: As Charles Fried has described it, traditional constitutional thinking "assumes intelligence and good faith" on the part of those who undertake it, and "that is a strength, not a weakness."[70] Our constitutional tradition provides the intellectual tools necessary to bring even the most intractable issues of constitutional governance under the rule of reason, while the tradition embodies moral ideals that can inspire the courage to act on what constitutional reason demands.

70. FRIED, *supra* note 1, at 67.

INDEX

www.ingramcontent.com/pod-product-compliance
Lightning Source LLC
Chambersburg PA
CBHW032226290126
39022CB00002B/33